INTEGRATED ACTIVE ANTENNAS AND SPATIAL POWER COMBINING • *Julio A. Navarro and Kai Chang*
FREQUENCY CONTROL OF SEMICONDUCTOR LASERS • *Motoichi Ohtsu (ed.)*
SOLAR CELLS AND THEIR APPLICATIONS • *Larry D. Partain (ed.)*
ANALYSIS OF MULTICONDUCTOR TRANSMISSION LINES • *Clayton R. Paul*
INTRODUCTION TO ELECTROMAGNETIC COMPATIBILITY • *Clayton R. Paul*
INTRODUCTION TO HIGH-SPEED ELECTRONICS AND OPTOELECTRONICS • *Leonard M. Riaziat*
NEW FRONTIERS IN MEDICAL DEVICE TECHNOLOGY • *Arye Rosen and Harel Rosen (eds.)*
ELECTROMAGNETIC PROPAGATION IN MULTI-MODE RANDOM MEDIA • *Harrison E. Rowe*
NONLINEAR OPTICS • *E. G. Sauter*
InP-BASED MATERIALS AND DEVICES: PHYSICS AND TECHNOLOGY • *Osamu Wada and Hideki Hasegawa (eds.)*
FREQUENCY SELECTIVE SURFACE AND GRID ARRAY • *T. K. Wu (ed.)*
ACTIVE AND QUASI-OPTICAL ARRAYS FOR SOLID-STATE POWER COMBINING • *Robert A. York and Zoya B. Popović (eds.)*
OPTICAL SIGNAL PROCESSING, COMPUTING AND NEURAL NETWORKS • *Francis T. S. Yu and Suganda Jutamulia*

```
QC448 .R69 1999
c.1
Rowe, H. E. (Harrison
E.)
Electromagnetic
propagation in
c1999.
```

Electromagnetic Propagation in Multi-Mode Random Media

Electromagnetic Propagation in Multi-Mode Random Media

HARRISON E. ROWE

A WILEY-INTERSCIENCE PUBLICATION
JOHN WILEY & SONS, INC.
NEW YORK / CHICHESTER / WEINHEIM / BRISBANE / SINGAPORE / TORONTO

This book is printed on acid-free paper. ∞

Copyright © 1999 by John Wiley & Sons, Inc. All rights reserved.

Published simultaneously in Canada.

No part of this publication may be reproduced, stored in a retrieval system or transmitted in any form or by any means, electronic, mechanical, photocopying, recording, scanning or otherwise, except as permitted under Sections 107 or 108 of the 1976 United States Copyright Act, without either the prior written permission of the Publisher, or authorization through payment of the appropriate per-copy fee to the Copyright Clearance Center, 222 Rosewood Drive, Danvers, MA 01923, (978) 750-8400, fax (978) 750-4744. Requests to the Publisher for permission should be addressed to the Permissions Department, John Wiley & Sons, Inc., 605 Third Avenue, New York, NY 10158-0012, (212) 850-6011, fax (212) 850-6008, E-Mail: PERMREQ@WILEY.COM.

Library of Congress Cataloging-in-Publication Data:

Rowe, H. E. (Harrison E.)
   Electromagnetic propagation in multi-mode random media / Harrison E. Rowe.
      p.   cm.
   Includes bibliographical references and index.
   ISBN 0-471-11003-5 (cloth : alk. paper)
   1. Fiber optics. 2. Optical wave guides. 3. Electromagnetic waves—Transmission. I. Title.
QC448.R69   1999
621.36'92—dc21   98-28787

Printed in the United States of America.

10 9 8 7 6 5 4 3 2 1

*To Alicia,
Amy, Elizabeth, Edward, and Alison,
and to the memory of
Stephen O. Rice*

Contents

1	**Introduction**	**1**
	References	3
2	**Coupled Line Equations**	**5**
	2.1 Introduction	5
	2.2 Two-Mode Coupled Line Equations	6
	2.3 Exact Solutions	8
	2.4 Discrete Approximation	11
	2.5 Perturbation Theory	13
	2.6 Multi-Mode Coupled Line Equations	15
	References	19
3	**Guides with White Random Coupling**	**21**
	3.1 Introduction	21
	3.2 Notation—Two-Mode Case	23
	3.3 Average Transfer Functions	25
	3.4 Coupled Power Equations	30
	3.5 Power Fluctuations	35
	3.6 Transfer Function Statistics	39
	3.7 Impulse Response Statistics	43
	3.8 Discussion	46
	References	47
4	**Examples—White Coupling**	**49**
	4.1 Introduction	49

	4.1.1 Single-Mode Input	50
	4.1.2 Multi-Mode Coherent Input	51
	4.1.3 Multi-Mode Incoherent Input	51
	4.2 Coupled Power Equations—Two-Mode Case	52
	4.2.1 Single-Mode Input	52
	4.2.2 Two-Mode Coherent Input	55
	4.2.3 Two-Mode Incoherent Input	58
	4.3 Power Fluctuations—Two-Mode Guide	58
	4.3.1 Single-Mode Input	60
	4.3.2 Two-Mode Coherent Input	60
	4.3.3 Two-Mode Incoherent Input	61
	4.3.4 Discussion	62
	4.4 Impulse Response—Two-Mode Case	64
	4.5 Coupled Power Equations—Four-Mode Case	67
	4.5.1 Single-Mode Input	69
	4.5.2 Multi-Mode Coherent Input	72
	4.5.3 Multi-Mode Incoherent Input	72
	4.6 Nondegenerate Case—Approximate Results	74
	4.6.1 Average Transfer Functions	75
	4.6.2 Coupled Power Equations	77
	4.6.3 Power Fluctuations	81
	4.6.4 Discussion	84
	4.7 Discussion	84
	References	85
5	**Directional Coupler with White Propagation Parameters**	**87**
	5.1 Introduction	87
	5.2 Statistical Model	88
	5.3 Average Transfer Functions	91
	5.4 Coupled Power Equations	93
	5.5 Discussion	95
	References	97
6	**Guides with General Coupling Spectra**	**99**
	6.1 Introduction	99
	6.2 Almost-White Coupling Spectra	100
	6.2.1 Two Modes	101

	6.2.2 N Modes	104
	6.3 General Coupling Spectra—Lossless Case	106
	6.3.1 Two Modes	109
	6.3.2 N Modes	110
	6.4 General Coupling Spectra—Lossy Case	115
	6.4.1 Two Modes	116
	6.4.2 N Modes	118
	6.5 Discussion	121
	References	122
7	**Four-Mode Guide with Exponential Coupling Covariance**	**123**
	7.1 Introduction	123
	7.2 Average Transfer Functions	126
	7.3 Coupled Power Equations	127
	7.4 Discussion	127
8	**Random Square-Wave Coupling**	**129**
	8.1 Introduction	129
	8.2 Two Modes—Binary Independent Sections	132
	8.3 Two Modes—Binary Markov Sections	135
	8.4 Four Modes—Multi-Level Markov Sections	138
	8.5 Discussion	143
9	**Multi-Layer Coatings with Random Optical Thickness**	**145**
	9.1 Introduction	145
	9.2 Matrix Analysis	147
	9.3 Kronecker Products	150
	9.4 Example: 13-Layer Filter	152
	9.4.1 Statistical Model	153
	9.4.2 Transmittance	155
	9.4.3 Two-Frequency Transmission Statistics	155
	9.5 Discussion	158
	References	159
10	**Conclusion**	**161**
	References	162

Appendix A	Series Solution for the Coupled Line Equations	**163**
	References	168
Appendix B	General Transmission Properties of Two-Mode Guide	**169**
	References	173
Appendix C	Kronecker Products	**175**
	References	177
Appendix D	Expected Values of Matrix Products	**179**
	D.1 Independent Matrices	179
	D.2 Markov Matrices	183
	D.2.1 Markov Chains	183
	D.2.2 Scalar Variables	184
	D.2.3 Markov Matrix Products	186
	References	190
Appendix E	Time- and Frequency-Domain Statistics	**191**
	E.1 Second-Order Impulse Response Statistics	191
	E.2 Time-Domain Analysis	195
	References	196
Appendix F	Symmetric Slab Waveguide—Lossless TE Modes	**197**
	F.1 General Results	197
	F.2 Example	200
	References	202
Appendix G	Equal Propagation Constants	**203**
Appendix H	Asymptotic Form of Coupled Power Equations	**209**
Appendix I	Differential Equations Corresponding to Matrix Equations	**211**
	I.1 Scalar Case	211

	I.2 Matrix Case	212
	References	214
Appendix J	**Random Square-Wave Coupling Statistics**	**215**
	J.1 Introduction	215
	J.2 Binary Sections	217
	J.2.1 Independent	218
	J.2.2 Markov	218
	J.3 Multi-Level Markov Sections	219
	J.3.1 Low-Pass—Six Levels	219
	J.3.2 Band-Pass—Five Levels	224
	References	226
Appendix K	**Matrix for a Multi-Layer Structure**	**227**
Index		**231**

Electromagnetic Propagation in Multi-Mode Random Media

CHAPTER ONE

Introduction

This text presents analytic methods for calculating the transmission statistics of microwave and optical components with random imperfections. Three general classes of devices are studied:

1. Multi-mode guides such as oversize waveguides or optical fibers.
2. Directional couplers.
3. Multi-layer optical coatings used as windows, mirrors, or filters, with plane-wave excitation.

All of these various transmission media and devices are multi-mode. For the first two categories the significant modes travel in the same direction, that is, forward; coupling to backward modes is neglected. In the third class of devices, the two modes are plane waves traveling in opposite directions.

Electromagnetic calculations yield equations that describe their performance in terms of coupling between modes. For multi-mode guides and directional couplers these are called the coupled line equations. For multi-layer coatings, they are the Fresnel reflection and transmission coefficients between adjacent layers. The starting point for all of our analyses will be these various equations; we assume their coefficients have been determined elsewhere by electromagnetic theory, in terms of the geometry and dielectric constants of the media comprising each device. No electromagnetic calculations are contained in the present work.

The performance of every such system is limited by random departures of its physical parameters from their ideal design values. These include both geometric and electrical parameters. Some examples are:

1. The axis of a multi-mode guide may exhibit random straightness deviations, or cross-sectional deformations such as slight ellipticity in a nominally circular guide.
2. A directional coupler made of two microstrip lines may show small random variations in the separation of the microstrip lines or random variations in their individual widths.
3. Microscopic dielectric constant variations may exist in the medium of either of the above two examples.
4. A multi-layer coating may have random errors in the optical thickness of the different layers, caused by variations in either the geometric thickness or the electrical parameters of the layers.

The statistics of the parameters in the corresponding coupled mode equations are determined by the statistics of these physical imperfections.

We characterize the transmission performance of each of these various systems by their complex transfer functions and the corresponding impulse responses. We determine the complex transfer function statistics as functions of the statistics of the coupling coefficients, propagation parameters, or the Fresnel coefficients appropriate to each case, and of the design parameters of the ideal system. These results in turn determine the corresponding time-domain statistics. The treatment is exclusively analytic; no Monte Carlo or other simulation methods are employed. Computer usage is restricted to symbolic operations, to evaluation of analytical expressions, and to creating plots.

The present text has the following goals:

1. Teaching the analytic methods.
2. Showing the different types of problems to which they may be applied.
3. Application to problems of significant practical interest.

Matrix techniques—in particular Kronecker products and related methods—play a central role in this work. Their application is natural for multi-layer devices. For continuous random coupling, as an intermediate step the guide is divided into statistically independent sections, each described by a wave matrix. In each case, this approach yields results with clarity and generality.

The present work has evolved from several early papers by the author and his colleagues [1–4]. These earlier results were obtained by pencil-and-paper analysis. The availability of computer programs capable of symbolic algebra, calculus, and matrix operations has greatly expanded the scope of these methods. MAPLE [5] has been used extensively throughout the present work. To the best of the author's knowledge many of the present results are new.

The main text is restricted to the analysis of transmission statistics of various classes of devices described above. A number of related topics are relegated to the appendices.

The present methods have application beyond the random mode coupling problems treated here. Kronecker products apply directly to the statistical analysis of any cascaded system characterized by a matrix product with random elements.

REFERENCES

1. Harrison E. Rowe and D. T. Young, "Transmission Distortion in Multimode Random Waveguides," *IEEE Transactions on Microwave Theory and Techniques*, Vol. MTT-20, June 1972, pp. 349–365.
2. D. T. Young and Harrison E. Rowe, "Optimum Coupling for Random Guides with Frequency-Dependent Coupling," *IEEE Transactions on Microwave Theory and Techniques*, Vol. MTT-20, June 1972, pp. 365–372.
3. Harrison E. Rowe and Iris M. Mack, "Coupled Modes with Random Propagation Constants," *Radio Science*, Vol. 16, July–August 1981, pp. 485–493.
4. Harrison E. Rowe, "Waves with Random Coupling and Random Propagation Constants," *Applied Scientific Research*, Vol. 41, 1984, pp. 237–255.
5. André Heck, *Introduction to Maple*, Springer-Verlag, New York, 1993.

CHAPTER TWO

Coupled Line Equations

2.1. INTRODUCTION

Optical and microwave transmission media of diverse physical forms have a common mathematical description in terms of the coupled line equations. Examples include oversize hollow metallic waveguide of various cross sections, dielectric waveguide, and optical fibers. Different frequency bands are appropriate to these various media, ranging from microwave and millimeter wave through optical frequencies. Single guides are used to carry signals from one place to another; pairs of similar guides comprise directional couplers.

An ideal guide, with constant geometry and material properties, transmits a set of modes that propagate independently of each other. A closed guide (e.g., a hollow metallic waveguide with perfectly conducting walls) supports an infinite discrete set of modes, of which a finite number are propagating; open structures (e.g., dielectric waveguides or optical fibers) have a finite number of discrete propagating modes plus a continuum of modes corresponding to the radiation field. We shall be concerned only with the propagating modes in the present work.

Random imperfections in these structures can arise from geometric or from material parameter departures from ideal design. Geometric imperfections include random straightness or cross-section variations; material imperfections arise from undesired dielectric constant, or index of refraction variations. Schelkunoff [1] observed that fields in an imperfect waveguide could be expressed as a sum over modes of the corresponding ideal waveguide. In the absence

6 COUPLED LINE EQUATIONS

of imperfections, the modes of an ideal guide are uncoupled, i.e., propagate independently; imperfections cause coupling between the modes. Directional couplers require intentional coupling between two guides.

The primary coupling in such structures occurs between modes traveling in the same direction; the coupling between modes traveling in opposite directions is normally small, and is neglected throughout the present work.

The coupled line equations serve as a common description for all of these media. The quantities in these equations that characterize the various transmission systems are the propagation parameters of and the coupling coefficients between the different propagating modes. These quantities may exhibit statistical variations arising from the geometric and material imperfections of the physical systems. Our task in following chapters is to determine transmission statistics in terms of coupling coefficient and/or propagation parameter statistics.

In this chapter, we examine the general properties and the deterministic solutions of the coupled line equations, that will be of use throughout the statistical treatment in several following chapters. We describe the two-mode and multi-mode cases separately; the analytical methods used in calculating transmission statistics are more clearly illustrated in the two-mode case, where expressions can be written out explicitly. Additional modes introduce only additional algebraic complexity, treated here by MAPLE without explicit presentation of intermediate results.

2.2. TWO-MODE COUPLED LINE EQUATIONS

The coupled line equations for two forward-traveling modes are [2–10]

$$I_0'(z) = -\Gamma_0(z)I_0(z) + jc(z)I_1(z),$$
$$I_1'(z) = jc(z)I_0(z) - \Gamma_1(z)I_1(z). \tag{2.1}$$

$$\Gamma_0(z) = \alpha_0(z) + j\beta_0(z), \qquad \Gamma_1(z) = \alpha_1(z) + j\beta_1(z). \tag{2.2}$$

The complex wave amplitude $I_i(z)$ is proportional to the transverse electric field of the ith mode at point z along the guide, normalized such that the power carried in this mode is given by

$$P_i(z) = |I_i(z)|^2, \quad i = 0, 1. \tag{2.3}$$

$\Gamma_0(z)$ and $\Gamma_1(z)$ represent the propagation parameters; the attenuation $\alpha_0(z), \alpha_1(z)$ and phase $\beta_0(z), \beta_1(z)$ are real and positive, and the coupling coefficient $c(z)$ is real. The form of coupling coefficient in Equation (2.1) is appropriate for systems whose elements dz possess geometric symmetry, e.g., guides with random straightness deviation and directional couplers. The Γ's and c are functions of the geometric and material parameters of the device, and of the frequency.

Let

$$P(z) = P_0(z) + P_1(z). \tag{2.4}$$

In the lossless case powers of different modes add, and $P(z)$ represents the total power in the guide. More generally, $P(z)$ is simply the sum of the mode powers computed individually. Substituting Equation (2.3) into (2.4), differentiating, and substituting Equation (2.1) for the resulting derivatives, we obtain

$$\frac{dP(z)}{dz} = -2\alpha_0(z)P_0(z) - 2\alpha_1(z)P_1(z). \tag{2.5}$$

This result states that each mode contributes to the decrease in $P(z)$ in proportion to the product of its attenuation constant and the power it carries. For the lossless case, $\alpha_0 = \alpha_1 = 0$, Equation (2.5) yields conservation of power.

Equations (2.1)–(2.2) approximate the response of a multi-mode guide in which only two forward-traveling modes are significant, and other modes may be neglected. We denote the signal (desired) mode by I_0 and the spurious (undesired) mode by I_1. For the ideal guide (without imperfections) Γ_0 and Γ_1 are constant, independent of z, and $c(z) = 0$. Since both modes travel in the forward $(+z)$ direction $\alpha_0, \alpha_1, \beta_0$, and β_1 are all positive. We assume that the signal mode has lower loss; $\alpha_0 \leq \alpha_1$. Random imperfections are modeled by regarding $c(z), \Gamma_0(z)$, and $\Gamma_1(z)$ as stationary random processes with appropriate statistics. As a particular example, a two-mode guide with random two-dimensional straightness deviations has constant propagation parameters Γ_0 and Γ_1 and a coupling coefficient inversely proportional to the radius of curvature $R(z)$ of the guide

8 COUPLED LINE EQUATIONS

axis,

$$c(z) = \frac{C}{R(z)}, \tag{2.6}$$

the constant C being a function of the two modes, determined by electromagnetic theory. In the special case of "single-mode" fiber, I_0 and I_1 represent the two polarizations of the dominant mode and $\Gamma_0 = \Gamma_1$.

The coupled line equations can describe a directional coupler where backward modes are neglected. For this case, random imperfections are modeled by taking the propagation parameters to be stationary random processes. As a simple example, $c(z)$ = constant and the propagation parameters have equal mean values, $\langle\Gamma_1(z)\rangle = \langle\Gamma_2(z)\rangle$.

Alternatively, Equations (2.1) and (2.2) can be applied to a transmission medium with a reflected wave. Denoting the forward wave by I_0 and the backward wave by I_1, $\Gamma_0 = -\Gamma_1 = \alpha + j\beta$ with positive α and β; c represents the reflection coefficient. However, we will not use these relations in our later treatment of multi-layer devices, but will rather employ direct methods.

Direct solution of the coupled line equations for arbitrary $c(z)$ and/or $\Gamma_0(z), \Gamma_1(z)$ is not possible. The matrix methods we employ to determine the transmission statistics for random guides require the solutions described in the remainder of this chapter.

2.3. EXACT SOLUTIONS

First, consider two degenerate forward-traveling modes, i.e., with equal propagation parameters:

$$\Gamma_0(z) = \Gamma_1(z) \equiv \Gamma(z). \tag{2.7}$$

The solutions to Equation (2.1) are

$$\begin{bmatrix} I_0(z) \\ I_1(z) \end{bmatrix} = e^{-\int_0^z \Gamma(x)dx} \begin{bmatrix} \cos\int_0^z c(x)dx & j\sin\int_0^z c(x)dx \\ j\sin\int_0^z c(x)dx & \cos\int_0^z c(x)dx \end{bmatrix} \cdot \begin{bmatrix} I_0(0) \\ I_1(0) \end{bmatrix}. \tag{2.8}$$

2.3. EXACT SOLUTIONS

In this case, only the integrals of $\Gamma(z)$ and $c(z)$ matter, their detailed functional behavior being unimportant. These results are a consequence of the fact that the two modes have identical attenuation and phase parameters; it doesn't matter how the coupling is distributed.

Equation (2.8) yields directly the wave matrix for delta-function coupling. Set

$$c(z) = c\delta(z), \tag{2.9}$$

where c is the magnitude of the delta function. Then,

$$\begin{bmatrix} I_0(0+) \\ I_1(0+) \end{bmatrix} = \mathbf{S} \cdot \begin{bmatrix} I_0(0) \\ I_1(0) \end{bmatrix}, \tag{2.10}$$

where the wave matrix for the discrete coupler is given by

$$\mathbf{S} = \begin{bmatrix} \cos c & j\sin c \\ j\sin c & \cos c \end{bmatrix}. \tag{2.11}$$

Comparison with Equation (2.6) shows that Equations (2.9)–(2.11) yield the response of a discrete tilt of angle c/C radians.

Next, consider the coupled line equations with constant coupling and constant propagation parameters, now called propagation constants:

$$c(z) = c_0; \quad \Gamma_0(z) = \Gamma_0, \quad \Gamma_1(z) = \Gamma_1. \tag{2.12}$$

Equations (2.1) become linear equations with constant coefficients, solved by elementary means [5]:

$$\begin{bmatrix} I_0(z) \\ I_1(z) \end{bmatrix} = \mathbf{T}(z) \cdot \begin{bmatrix} I_0(0) \\ I_1(0) \end{bmatrix}. \tag{2.13}$$

$$\mathbf{T}(z) = \left[e^{-\frac{\Gamma_0+\Gamma_1}{2}z} / (K_+ - K_-) \right]$$
$$\cdot \begin{bmatrix} -K_- e^{(\Delta\Gamma/2)z\sqrt{}} + K_+ e^{-(\Delta\Gamma/2)z\sqrt{}} & e^{(\Delta\Gamma/2)z\sqrt{}} - e^{-(\Delta\Gamma/2)z\sqrt{}} \\ e^{(\Delta\Gamma/2)z\sqrt{}} - e^{-(\Delta\Gamma/2)z\sqrt{}} & K_+ e^{(\Delta\Gamma/2)z\sqrt{}} - K_- e^{-(\Delta\Gamma/2)z\sqrt{}} \end{bmatrix}.$$

$$\tag{2.14}$$

COUPLED LINE EQUATIONS

$$K_{\pm} = -j\frac{1 \pm \sqrt{}}{2c_0/\Delta\Gamma}; \qquad K_+ K_- = -1. \tag{2.15}$$

$$\frac{1}{K_+ - K_-} = j\frac{c_0/\Delta\Gamma}{\sqrt{}}. \tag{2.16}$$

$$\sqrt{} = \sqrt{1 - (2c_0/\Delta\Gamma)^2}. \tag{2.17}$$

$$\Delta\Gamma = \Gamma_0 - \Gamma_1. \tag{2.18}$$

Let us approach a discrete coupler, Equation (2.9), by setting

$$c_0 = \frac{c}{z} \tag{2.19}$$

with fixed c, and taking the limit as $z \to 0$ (and $c_0 \to \infty$) in Equations (2.12)–(2.18). Then [6],

$$\lim_{z \to 0} (\Delta\Gamma/2)z\sqrt{1 - [2c/(\Delta\Gamma z)]^2} = jc. \tag{2.20}$$

$$\lim_{z \to 0} K_{\pm} = \pm 1. \tag{2.21}$$

$$\lim_{z \to 0} \frac{1}{K_+ - K_-} = \frac{1}{2}. \tag{2.22}$$

$$\lim_{z \to 0} e^{-\frac{\Gamma_0 + \Gamma_1}{2} z} = 1. \tag{2.23}$$

Substituting Equations (2.20)–(2.23) into Equations (2.14)–(2.17), we obtain the limiting form of $T(z)$:

$$\lim_{z \to 0} \mathbf{T}(z) = \begin{bmatrix} \cos c & j \sin c \\ j \sin c & \cos c \end{bmatrix}. \tag{2.24}$$

Equation (2.24) is identical to Equation (2.11). This result has the following physical interpretation. As the coupling becomes larger over a shorter length of guide, in such a way that the integrated coupling remains constant, the length becomes so short that the differential attenuation and phase shift between the two modes becomes negligible. Moreover, this argument applies not only for constant Γ_0 and Γ_1 as in Equation (2.12), but holds more generally for propagation parameters with arbitrary functional dependence. Hence, the result of Equations (2.11) and (2.24) holds in general for a discrete coupler.

2.4. DISCRETE APPROXIMATION

Next, we determine a discrete approximation to a guide with continuous coupling [7]. This is necessary to apply matrix techniques in the calculation of transmission statistics for random guides. Divide the guide into sections Δz long. The discussion of Section 2.3 suggests that for small enough Δz we can lump all the coupling in each section at the end of the section; i.e., we may replace $c(z)$ by $c_\delta(z)$ as follows:

$$c_\delta(z) = \sum_{k=1} c_k \delta(z - k\Delta z), \qquad (2.25)$$

where

$$c_k = \int_{(k-1)\Delta z}^{k\Delta z} c(z)dz. \qquad (2.26)$$

The kth section is illustrated in Figure 2.1, which shows $c(z)$ and $c_\delta(z)$; this approximation places all of the coupling at the end of the section.

The corresponding matrix description becomes

$$\begin{bmatrix} I_0(k\Delta z) \\ I_1(k\Delta z) \end{bmatrix} = \begin{bmatrix} \cos c_k & j\sin c_k \\ j\sin c_k & \cos c_k \end{bmatrix} \cdot \begin{bmatrix} e^{-\gamma_{0k}} & 0 \\ 0 & e^{-\gamma_{1k}} \end{bmatrix} \cdot \begin{bmatrix} I_0[(k-1)\Delta z] \\ I_1[(k-1)\Delta z] \end{bmatrix}, \qquad (2.27)$$

where c_k is given by Equation (2.26) and the γ_k's give the total attenuation and phase shift in the length Δz:

$$\begin{aligned} \gamma_{0k} &= \int_{(k-1)\Delta z}^{k\Delta z} \Gamma_0(z)dz = \int_{(k-1)\Delta z}^{k\Delta z} \alpha_0(z)dz + j\int_{(k-1)\Delta z}^{k\Delta z} \beta_0(z)dz. \\ \gamma_{1k} &= \int_{(k-1)\Delta z}^{k\Delta z} \Gamma_1(z)dz = \int_{(k-1)\Delta z}^{k\Delta z} \alpha_1(z)dz + j\int_{(k-1)\Delta z}^{k\Delta z} \beta_1(z)dz. \end{aligned} \qquad (2.28)$$

Equations (2.26)–(2.28) will approximate the solutions to the coupled line equations if the differential attenuation and phase shift are

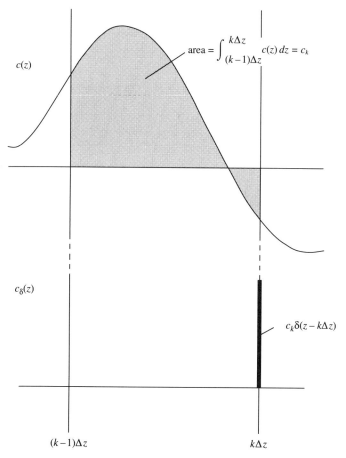

FIGURE 2.1. Discrete approximation for a guide section.

small in any section of guide of length less than Δz:

$$\left| \int_{z_1}^{z_2} [\alpha_0(z) - \alpha_1(z)] dz \right| \ll 1,$$
$$\left| \int_{z_1}^{z_2} [\beta_0(z) - \beta_1(z)] dz \right| \ll 2\pi, \qquad 0 < z_2 - z_1 \leq \Delta z. \quad (2.29)$$

If $\alpha(z)$ and $\beta(z)$ are constant, these restrictions become

$$\Delta z \ll \frac{1}{\alpha_0 - \alpha_1}, \qquad \Delta z \ll \frac{2\pi}{|\beta_0 - \beta_1|}. \quad (2.30)$$

Note that no small-coupling assumptions have been necessary here. Subject to Equation (2.29) or (2.30), the matrix result of Equations (2.26)–(2.28) will yield a good approximation to the solutions of Equations (2.1) and (2.2).

We will subsequently need to divide the guide into sections Δz that are approximately statistically independent. Equations (2.26)–(2.28) are appropriate for this purpose when Δz satisfying Equation (2.29) or (2.30) is long compared to the correlation length of the random coupling or propagation parameters, i.e., for random parameters having white or almost white spectra.

2.5. PERTURBATION THEORY

For non-white coupling or propagation parameter spectra, Δz satisfying Equation (2.29) or (2.30) will be short compared to the correlation length of the random parameters. In this case, Equations (2.25)–(2.28) are replaced by results based on perturbation theory to describe the response of a guide section long compared to the correlation length, for small coupling [2–4, 6]. It is convenient to normalize the complex wave amplitudes as follows:

$$I_0(z) = e^{-\gamma_0(z)} G_0(z),$$
$$I_1(z) = e^{-\gamma_1(z)} G_1(z), \tag{2.31}$$

where

$$\gamma_0(z) = \int_0^z \Gamma_0(x)dx, \qquad \gamma_1(z) = \int_0^z \Gamma_1(x)dx. \tag{2.32}$$

Then substituting into Equation (2.1) we find

$$G_0'(z) = jc(z)e^{+\Delta\gamma(z)} G_1(z),$$
$$G_1'(z) = jc(z)e^{-\Delta\gamma(z)} G_0(z), \tag{2.33}$$

where

$$\Delta\gamma(z) = \int_0^z \Delta\Gamma(x)dx, \qquad \Delta\Gamma(z) = \Gamma_0(z) - \Gamma_1(z). \tag{2.34}$$

14 COUPLED LINE EQUATIONS

Then the method of successive approximations given in Appendix A yields the following approximate results:

$$\begin{bmatrix} G_0(z) \\ G_1(z) \end{bmatrix} \approx \mathbf{M} \cdot \begin{bmatrix} G_0(0) \\ G_1(0) \end{bmatrix} = \begin{bmatrix} m_{00} & m_{01} \\ m_{10} & m_{11} \end{bmatrix} \cdot \begin{bmatrix} G_0(0) \\ G_1(0) \end{bmatrix}, \quad (2.35)$$

where the elements of $\mathbf{M}$ are given as follows:

$$\begin{aligned} m_{00} &= 1 - \int_0^z c(x)e^{+\Delta\gamma(x)}dx \int_0^x c(y)e^{-\Delta\gamma(y)}dy. \\ m_{01} &= j\int_0^z c(x)e^{+\Delta\gamma(x)}dx. \\ m_{10} &= j\int_0^z c(x)e^{-\Delta\gamma(x)}dx. \\ m_{11} &= 1 - \int_0^z c(x)e^{-\Delta\gamma(x)}dx \int_0^x c(y)e^{+\Delta\gamma(y)}dy. \end{aligned} \quad (2.36)$$

For constant propagation parameters,

$$\Gamma_0(z) = \Gamma_0, \qquad \Gamma_1(z) = \Gamma_1, \qquad \Delta\Gamma(z) = \Delta\Gamma = \Gamma_0 - \Gamma_1, \quad (2.37)$$

these results simplify by the substitution

$$\Delta\gamma(z) \to \Delta\Gamma z. \quad (2.38)$$

In this case, two alternative forms for m_{00} and m_{11} are as follows:

$$\begin{aligned} m_{00} &= 1 - \int_0^z c(x)e^{+\Delta\Gamma x}dx \int_0^x c(y)e^{-\Delta\Gamma y}dy \\ &= 1 - \int_0^z e^{+\Delta\Gamma \zeta}d\zeta \int_0^{z-\zeta} c(x)c(x+\zeta)dx \\ &= 1 - \frac{1}{2}\int_0^z \int_0^z c(x)c(y)e^{+\Delta\Gamma|x-y|}dxdy. \end{aligned} \quad (2.39)$$

$$\begin{aligned} m_{11} &= 1 - \int_0^z c(x)e^{-\Delta\Gamma x}dx \int_0^x c(y)e^{+\Delta\Gamma y}dy \\ &= 1 - \int_0^z e^{-\Delta\Gamma \zeta}d\zeta \int_0^{z-\zeta} c(x)c(x+\zeta)dx \\ &= 1 - \frac{1}{2}\int_0^z \int_0^z c(x)c(y)e^{-\Delta\Gamma|x-y|}dxdy. \end{aligned} \quad (2.40)$$

Each of these three expressions for m_{00} and for m_{11} has its uses. The first yields a physical interpretation of these perturbation results. The expected value of the middle expressions introduces the covariance of the coupling coefficient for guides with random coupling, and constant propagation parameters.

Equations (2.35)–(2.40) yield useful approximations for small coupling. They are the first terms of infinite series given in Appendix A. These series converge rapidly when

$$\int_0^z |c(x)|dx \ll 1. \tag{2.41}$$

2.6. MULTI-MODE COUPLED LINE EQUATIONS

The above relations for two modes are readily extended to many forward modes [2–6] using matrix notation as follows. Equations (2.1) and (2.2), the coupled line equations, become

$$I'(z) = -\Gamma(z) \cdot I(z) + jc(z)\mathbf{C} \cdot I(z), \tag{2.42}$$

where

$$I^T(z) = \begin{bmatrix} I_0(z) & I_1(z) & I_2(z) & \cdots \end{bmatrix}, \tag{2.43}$$

$$\Gamma(z) = \begin{bmatrix} \Gamma_0(z) & 0 & 0 & \cdots \\ 0 & \Gamma_1(z) & 0 & \cdots \\ 0 & 0 & \Gamma_2(z) & \cdots \\ \vdots & \vdots & \vdots & \ddots \end{bmatrix}, \tag{2.44}$$

$$\mathbf{C} = \begin{bmatrix} 0 & C_{01} & C_{02} & \cdots \\ C_{01} & 0 & C_{12} & \cdots \\ C_{02} & C_{12} & 0 & \cdots \\ \vdots & \vdots & \vdots & \ddots \end{bmatrix}, \quad \mathbf{C} = \mathbf{C}^T = \mathbf{C}^*. \tag{2.45}$$

The mode powers are

$$P_i(z) = |I_i(z)|^2. \tag{2.46}$$

16 COUPLED LINE EQUATIONS

Define

$$P(z) = \sum_i P_i(z) = \mathbf{I}^{\mathrm{T}}(z) \cdot \mathbf{I}^*(z). \tag{2.47}$$

Then,

$$\frac{dP(z)}{dz} = \mathbf{I}^{\mathrm{T}}(z) \cdot \mathbf{I}'^*(z) + \mathbf{I}'^{\mathrm{T}}(z) \cdot \mathbf{I}^*(z). \tag{2.48}$$

Substituting Equation (2.42) into (2.48), the multi-mode generalization of Equation (2.5) becomes

$$\frac{dP(z)}{dz} = -\mathbf{I}^{\mathrm{T}}(z) \cdot \{\mathbf{\Gamma}(z) + \mathbf{\Gamma}^*(z)\} \cdot \mathbf{I}^*(z) = -2 \sum_i \alpha_i(z) P_i(z). \tag{2.49}$$

For the lossless case all $\alpha_i = 0$, and Equation (2.49) yields conservation of power.

Exact solutions corresponding to those of Section 2.3 for two modes are readily found for the multi-mode case. In the degenerate case, all modes have identical propagation constants,

$$\Gamma(z) = \Gamma_0(z) = \Gamma_1(z) = \Gamma_2(z) = \cdots; \tag{2.50}$$

i.e., in matrix notation

$$\mathbf{\Gamma}(z) = \Gamma(z)\mathcal{I}, \tag{2.51}$$

where $\mathcal{I}$ represents the unit matrix:

$$\mathcal{I} = \begin{bmatrix} 1 & 0 & 0 & \cdots \\ 0 & 1 & 0 & \cdots \\ 0 & 0 & 1 & \cdots \\ \vdots & \vdots & \vdots & \ddots \end{bmatrix}. \tag{2.52}$$

The solution to Equation (2.42) for this case is

$$\mathbf{I}(z) = e^{-\int_0^z \Gamma(x)dx} e^{j\int_0^z c(x)dx \cdot \mathbf{C}} \cdot \mathbf{I}(0). \tag{2.53}$$

For delta-function coupling, set

$$c(z) = c\delta(z) \tag{2.54}$$

2.6. MULTI-MODE COUPLED LINE EQUATIONS

in Equation (2.53). Then,

$$I(0+) = S \cdot I(0), \qquad (2.55)$$

where the wave matrix for the multi-mode discrete coupler is given by

$$S = e^{jcC}. \qquad (2.56)$$

Finally, for constant coupling and constant propagation parameters Equation (2.42) becomes

$$I'(z) = -\Gamma \cdot I(z) + jc_0 C \cdot I(z), \qquad (2.57)$$

where Γ is the constant matrix

$$\Gamma = \begin{bmatrix} \Gamma_0 & 0 & 0 & \cdots \\ 0 & \Gamma_1 & 0 & \cdots \\ 0 & 0 & \Gamma_2 & \cdots \\ \vdots & \vdots & \vdots & \ddots \end{bmatrix}. \qquad (2.58)$$

The solution to Equations (2.57) and (2.58) in matrix notation is

$$I(z) = e^{\{-\Gamma + jc_0 C\} z} \cdot I(0). \qquad (2.59)$$

The exact multi-mode solutions given in Equations (2.50)–(2.59) specialize to those for the two-mode case of Equations (2.7)–(2.24) by evaluating the matrix exponentials. We perform this reduction for the discrete coupler. Set

$$cC = \begin{bmatrix} 0 & c \\ c & 0 \end{bmatrix} \qquad (2.60)$$

in Equation (2.56). Diagonalizing this matrix,

$$\begin{bmatrix} 0 & c \\ c & 0 \end{bmatrix} = \begin{bmatrix} 1 & 1 \\ 1 & -1 \end{bmatrix} \cdot \begin{bmatrix} c & 0 \\ 0 & -c \end{bmatrix} \cdot \begin{bmatrix} 1 & 1 \\ 1 & -1 \end{bmatrix}^{-1}$$

$$= \frac{1}{2} \begin{bmatrix} 1 & 1 \\ 1 & -1 \end{bmatrix} \cdot \begin{bmatrix} c & 0 \\ 0 & -c \end{bmatrix} \cdot \begin{bmatrix} 1 & 1 \\ 1 & -1 \end{bmatrix}. \qquad (2.61)$$

Then Equation (2.56) yields

$$e^{jcC} = \frac{1}{2}\begin{bmatrix} 1 & 1 \\ 1 & -1 \end{bmatrix} \cdot \begin{bmatrix} e^{jc} & 0 \\ 0 & e^{-jc} \end{bmatrix} \cdot \begin{bmatrix} 1 & 1 \\ 1 & -1 \end{bmatrix} = \begin{bmatrix} \cos c & j\sin c \\ j\sin c & \cos c \end{bmatrix},$$
(2.62)

in agreement with Equation (2.11).

The discrete approximation of Section 2.4, used subsequently for white or almost white coupling $c(z)$, generalizes to the multi-mode case as follows:

$$I(k\Delta z) = e^{jc_k C} \cdot e^{-\gamma_k(z)} \cdot I[(k-1)\Delta z], \qquad (2.63)$$

where as in Equation (2.26),

$$c_k = \int_{(k-1)\Delta z}^{k\Delta z} c(z)dz, \qquad (2.64)$$

and

$$\gamma_k = \int_{(k-1)\Delta z}^{k\Delta z} \Gamma(z)dz. \qquad (2.65)$$

The restrictions of Equation (2.29) generalize to

$$\left| \int_{z_1}^{z_2} [\alpha_i(z) - \alpha_j(z)]dz \right| \ll 1,$$
$$\left| \int_{z_1}^{z_2} [\beta_i(z) - \beta_j(z)]dz \right| \ll 2\pi, \qquad 0 < z_2 - z_1 \leq \Delta z; \quad i, j = 0, 1, 2, \cdots.$$
(2.66)

Finally, for multi-mode perturbation theory

$$I(z) = e^{-\gamma(z)} \cdot G(z), \qquad (2.67)$$

where

$$\gamma(z) = \int_0^z \Gamma(x)dx. \qquad (2.68)$$

Substituting into Equation (2.42), we find

$$G'(z) = jc(z) \cdot e^{+\gamma(z)} \cdot C \cdot e^{-\gamma(z)} \cdot G(z) \qquad (2.69)$$

as the generalization of Equation (2.33). The method of successive approximations in Appendix A yields the following:

$$G(z) \approx \left\{ \mathcal{I} + j \int_0^z c(x) e^{+\gamma(x)} \cdot \mathbf{C} \cdot e^{-\gamma(x)} dx \right.$$
$$- \int_0^z c(x) e^{+\gamma(x)} \cdot \mathbf{C} \cdot e^{-\gamma(x)} dx$$
$$\left. \cdot \int_0^x c(y) e^{+\gamma(y)} \cdot \mathbf{C} \cdot e^{-\gamma(y)} dy \right\} \cdot G(0). \qquad (2.70)$$

This result holds for small coupling, i.e., from Equations (A.22) and (A.24)

$$\int_0^z |c(x)| dx \|\mathbf{C}\| \ll 1, \qquad (2.71)$$

where the matrix norm $\|\mathbf{C}\|$ is given by Equation (A.25). Equation (2.70) is required for the non-white case, where the correlation lengths of $c(z)$ or of the $\Gamma_i(z)$ are long compared to Δz satisfying Equation (2.66).

Finally, substituting

$$c(z)\mathbf{C} = \begin{bmatrix} 0 & c(z) \\ c(z) & 0 \end{bmatrix} \qquad (2.72)$$

and

$$\gamma(z) = \begin{bmatrix} \int_0^z \Gamma_0(x) dx & 0 \\ 0 & \int_0^z \Gamma_1(x) dx \end{bmatrix} \qquad (2.73)$$

into Equation (2.70) specializes this result to Equations (2.35) and (2.36) for the two-mode case.

REFERENCES

1. S. A. Schelkunoff, "Conversion of Maxwell's Equations into Generalized Telegraphist's Equations," *Bell System Technical Journal*, Vol. 34, September 1955, pp. 995–1043.

2. Huang Hung-chia, *Coupled Mode Theory*, VNU Science Press, Utrecht, The Netherlands, 1984.
3. Dietrich Marcuse, *Light Transmission Optics*, 2nd ed., Robert E. Krieger, Malabar, FL, 1989.
4. Dietrich Marcuse, *Theory of Dielectric Optical Waveguides*, 2nd ed., Academic Press, New York, 1991.
5. S. E. Miller, "Coupled Wave Theory and Waveguide Applications," *Bell System Technical Journal*, Vol. 33, May 1954, pp. 661–719.
6. H. E. Rowe and W. D. Warters, "Transmission in Multimode Waveguide with Random Imperfections," *Bell System Technical Journal*, Vol. 41, May 1962, pp. 1031–1170.
7. Harrison E. Rowe and D. T. Young, "Transmission Distortion in Multimode Random Waveguides," *IEEE Transactions on Microwave Theory and Techniques*, Vol. MTT-20, June 1972, pp. 349–365.
8. Harrison E. Rowe, "Waves with Random Coupling and Random Propagation Constants," *Applied Scientific Research*, Vol. 41, 1984, pp. 237–255.
9. Hermann A. Haus and Weiping Huang, "Coupled-Mode Theory," *Proceedings of the IEEE*, Vol. 79, October 1991, pp. 1505–1518.
10. Wei-Ping Huang, "Coupled-mode theory for optical waveguides: an overview," *Journal of the Optical Society of America A*, Vol. 11, March 1994, pp. 963–983.

CHAPTER THREE

Guides with White Random Coupling

3.1. INTRODUCTION

We consider the coupled line equations of Chapter 2 with white coupling coefficient and constant propagation parameters in the present chapter. This case is significant in that exact results for transmission statistics are obtained. Assume the coupling coefficient $c(z)$ of Chapter 2 is a zero-mean stationary random process with delta-function covariance

$$R_c(\zeta) = \langle c(z+\zeta)c(z)\rangle = S_0\delta(\zeta) \qquad (3.1)$$

and white spectrum

$$S(\nu) = \int_{-\infty}^{\infty} R_c(\zeta)e^{-j2\pi\nu\zeta}d\zeta = S_0, \qquad (3.2)$$

with spectral density S_0. We require the following additional property for every pair of nonoverlapping intervals:

$\int_{z1}^{z2} c(x)dx$ and $\int_{z3}^{z4} c(x)dx$ are statistically independent for

$$z1 < z2 < z3 < z4, \qquad (3.3)$$

i.e., different sections of guide have statistically independent coupling. In some cases, e.g., Gaussian or Poisson $c(z)$, Equations (3.1)

22 GUIDES WITH WHITE RANDOM COUPLING

and (3.2) imply Equation (3.3). We require Equation (3.3) in any case; it implies that $c(z)$ has a white spectrum.

For constant propagation parameters, Equations (2.1) and (2.2) for two modes become

$$I'_0(z) = -\Gamma_0 I_0(z) + jc(z)I_1(z),$$
$$I'_1(z) = jc(z)I_0(z) - \Gamma_1 I_1(z). \tag{3.4}$$

$$\Gamma_0 = \alpha_0 + j\beta_0, \qquad \Gamma_1 = \alpha_1 + j\beta_1. \tag{3.5}$$

Γ_0 and Γ_1 are independent of z. The normalized relations of Equations (2.31)–(2.34) become

$$I_0(z) = e^{-\Gamma_0 z} G_0(z),$$
$$I_1(z) = e^{-\Gamma_1 z} G_1(z). \tag{3.6}$$

$$G'_0(z) = jc(z)e^{+\Delta\Gamma z} G_1(z),$$
$$G'_1(z) = jc(z)e^{-\Delta\Gamma z} G_0(z). \tag{3.7}$$

$$\Delta\Gamma = \Gamma_0 - \Gamma_1 = \Delta\alpha + j\Delta\beta,$$
$$\Delta\alpha = \alpha_0 - \alpha_1, \qquad \Delta\beta = \beta_0 - \beta_1. \tag{3.8}$$

The corresponding results for the multi-mode case are obtained by substituting Equation (2.58) into Equations (2.42) and (2.67)–(2.69):

$$\mathbf{I}'(z) = -\mathbf{\Gamma} \cdot \mathbf{I}(z) + jc(z)\mathbf{C} \cdot \mathbf{I}(z). \tag{3.9}$$

$$\mathbf{\Gamma} = \begin{bmatrix} \Gamma_0 & 0 & 0 & \cdots \\ 0 & \Gamma_1 & 0 & \cdots \\ 0 & 0 & \Gamma_2 & \cdots \\ \vdots & \vdots & \vdots & \ddots \end{bmatrix}. \tag{3.10}$$

$$\mathbf{I}(z) = e^{-\mathbf{\Gamma} z} \cdot \mathbf{G}(z). \tag{3.11}$$

$$\mathbf{G}'(z) = jc(z) \cdot e^{+\mathbf{\Gamma} z} \cdot \mathbf{C} \cdot e^{-\mathbf{\Gamma} z} \cdot \mathbf{G}(z). \tag{3.12}$$

$$\mathbf{C} = \begin{bmatrix} 0 & C_{01} & C_{02} & \cdots \\ C_{01} & 0 & C_{12} & \cdots \\ C_{02} & C_{12} & 0 & \cdots \\ \vdots & \vdots & \vdots & \ddots \end{bmatrix}, \qquad \mathbf{C} = \mathbf{C}^T = \mathbf{C}^*. \qquad (3.13)$$

While the Γ_i are independent of z, they are frequency dependent; the coupling between modes, a random function of z, may also be frequency dependent. The G_i are thus functions of z and of frequency f. The frequency dependence is implicit in the above equations, and in subsequent solutions for single-frequency transmission statistics, such as average mode powers. However, additional notation is required to display the frequency dependence explicitly in the analysis of transfer-function frequency- and time-response statistics.

3.2. NOTATION—TWO-MODE CASE

We consider the solution to Equation (3.7) for the normalized signal transfer function G_0 for some fixed length $z = L$, with initial conditions

$$G_0(0) = 1, \qquad G_1(0) = 0. \qquad (3.14)$$

A unit sinusoidal signal is input at $z = 0$; the spurious mode is zero at the input. $G_0(L)$ and $G_1(L)$ are the normalized signal and signal-spurious mode transfer functions, respectively.

It is natural to suppress the L dependence, and instead display the frequency dependence. We first separate the z and f dependence of the coupling. The coupling coefficient $c(z)$ is proportional to some geometric parameter, which we denote by $d(z)$. For the two-mode case,

$$c(z) = Cd(z). \qquad (3.15)$$

$d(z)$ does not depend on f, but is a function only of z; any frequency dependence is contained in the parameter C. In the example of Equation (2.6), $d(z)$ is the curvature of the guide axis; in other applications it might be ellipticity, etc. For the multi-mode case, we may absorb the frequency dependence into the elements of the matrix **C** of Equation (3.13), and regard $c(z)$ as the frequency-independent geometric parameter.

Let us now consider the solution to Equation (3.7) for the normalized signal transfer function G_0, for some fixed length L and a given geometric parameter $d(z)$. G_0 is a function of $\Delta\alpha$, $\Delta\beta$, and C, and through these parameters a function of frequency f. However, the principal frequency dependence will normally occur through $\Delta\beta$. This suggests investigating the properties of G_0 as a function of $\Delta\beta$, with $\Delta\alpha$ and C regarded as fixed. Toward this end, we define the Fourier transform of G_0 as follows [1]:

$$g_{\Delta\alpha}(\tau) = \int_{-\infty}^{\infty} G_0(\Delta\alpha, \Delta\beta) e^{-j2\pi\tau(\frac{\Delta\beta L}{2\pi})} d\left(\frac{\Delta\beta L}{2\pi}\right)$$

$$= \frac{1}{2\pi} \int_{-\infty}^{\infty} G_0(\Delta\alpha, \Delta\beta) e^{-j\tau\Delta\beta L} d(\Delta\beta L). \qquad (3.16)$$

The inverse transform is

$$G_0(\Delta\alpha, \Delta\beta) = \int_{-\infty}^{\infty} g_{\Delta\alpha}(\tau) e^{+j\tau\Delta\beta L} d\tau. \qquad (3.17)$$

The fixed parameters L, $d(z)$, and C are suppressed in these two relations.

Consider an idealized guide in which $\Delta\alpha$ and C are frequency independent and $\Delta\beta$ is strictly linear with f. In a real guide $\Delta\alpha$ is an even function of f, $\Delta\beta$ and C odd. Therefore for the idealized guide C must satisfy

$$C = C_0 \operatorname{sgn} f = \begin{cases} |C_0|, & f > 0, \\ -C_0, & f < 0, \end{cases} \quad C_0 \text{ a positive constant.} \qquad (3.18)$$

The guide is dispersionless:

$$\Delta\beta = -2\pi f \left(\frac{1}{v_1} - \frac{1}{v_0}\right), \quad v_0 > v_1, \qquad (3.19)$$

where v_0 and v_1 are the signal and spurious mode velocities, respectively. Let

$$\Delta T = \left(\frac{1}{v_1} - \frac{1}{v_0}\right) L \qquad (3.20)$$

be the differential delay between signal and spurious mode transit time over length L. Then substituting in Equation (3.16),

$$g_{\Delta\alpha}(\tau) = \Delta T \int_{-\infty}^{\infty} G_0(\Delta\alpha, \Delta\beta) e^{j2\pi f \tau \cdot \Delta T} df. \quad (3.21)$$

Denote the signal mode impulse response by $\mathscr{g}_{\Delta\alpha}(t)$:

$$\mathscr{g}_{\Delta\alpha}(t) = \int_{-\infty}^{\infty} G_0(\Delta\alpha, \Delta\beta) e^{j2\pi f t} df. \quad (3.22)$$

Comparison with Equation (3.21) shows that

$$\mathscr{g}_{\Delta\alpha}(t) = \frac{1}{\Delta T} g_{\Delta\alpha}\left(\frac{t}{\Delta T}\right). \quad (3.23)$$

$g_{\Delta\alpha}(\tau)$ of Equations (3.16) and (3.17) is therefore the normalized impulse response of the idealized dispersionless guide with frequency-independent attenuation constants and coupling.

While Equation (3.23) is strictly true only for an idealized guide, it will provide a useful approximation for a physical guide in many cases. This approximation requires the frequency variation of $\Delta\alpha$, C, and of the group delay of the two modes to be small over the frequency range of interest.

The general properties of $G_0(\alpha, \beta)$ and $g_{\Delta\alpha}(\tau)$ are described in Appendix B. These properties are reflected in the statistical results to follow.

3.3. AVERAGE TRANSFER FUNCTIONS

For the two-mode case with constant propagation parameters, Equation (2.27) becomes

$$\begin{bmatrix} I_0(k\Delta z) \\ I_1(k\Delta z) \end{bmatrix} = \begin{bmatrix} e^{-\Gamma_0 \Delta z} \cos c_k & e^{-\Gamma_1 \Delta z} j \sin c_k \\ e^{-\Gamma_0 \Delta z} j \sin c_k & e^{-\Gamma_1 \Delta z} \cos c_k \end{bmatrix} \cdot \begin{bmatrix} I_0[(k-1)\Delta z] \\ I_1[(k-1)\Delta z] \end{bmatrix}, \quad (3.24)$$

where c_k is given by Equation (2.26) and illustrated in Figure 2.1. This approximation requires that the restrictions of Equation (2.30) are satisfied:

$$\Delta z \ll \frac{1}{|\Delta\alpha|}, \qquad \Delta z \ll \frac{2\pi}{|\Delta\beta|}. \tag{3.25}$$

By Equations (3.1) and (3.2), $c(z)$ is white with zero mean; by Equation (3.3), the different sections remain independent for any Δz. Therefore Equation (3.24) may be identified with Equation (D.8). Define the average complex mode amplitudes as

$$\mathcal{I}_i(z) = \langle I_i(z) \rangle, \quad \mathcal{G}_i(z) = \langle G_i(z) \rangle; \tag{3.26}$$

by Equation (D.14), the expected value of Equation 3.24 yields

$$\begin{bmatrix} \mathcal{I}_0(k\Delta z) \\ \mathcal{I}_1(k\Delta z) \end{bmatrix} = \begin{bmatrix} e^{-\Gamma_0 \Delta z} \langle \cos c_k \rangle & e^{-\Gamma_1 \Delta z} j \langle \sin c_k \rangle \\ e^{-\Gamma_0 \Delta z} j \langle \sin c_k \rangle & e^{-\Gamma_1 \Delta z} \langle \cos c_k \rangle \end{bmatrix} \cdot \begin{bmatrix} \mathcal{I}_0[(k-1)\Delta z] \\ \mathcal{I}_1[(k-1)\Delta z] \end{bmatrix}. \tag{3.27}$$

For sufficiently small Δz,[1]

$$\langle \cos c_k \rangle \approx 1 - \frac{1}{2}\langle c_k^2 \rangle = 1 - \frac{1}{2}S_0 \Delta z,$$
$$\langle \sin c_k \rangle \approx \langle c_k \rangle = 0. \tag{3.28}$$

Substituting Equation (3.28) into Equation (3.27) and taking the limit as $\Delta z \to 0$,

$$\mathcal{I}'_0(z) = -\left(\Gamma_0 + \frac{S_0}{2}\right)\mathcal{I}_0(z),$$
$$\mathcal{I}'_1(z) = -\left(\Gamma_1 + \frac{S_0}{2}\right)\mathcal{I}_1(z). \tag{3.29}$$

[1] From Equation (2.26), since $c(z)$ has zero mean,

$$\langle c_k \rangle = \int_{(k-1)\Delta z}^{k\Delta z} \langle c(z) \rangle \, dz = 0.$$

Using Equations (2.26) and (3.1),

$$\langle c_k^2 \rangle = \int_{(k-1)\Delta z}^{k\Delta z} \int_{(k-1)\Delta z}^{k\Delta z} \langle c(z)c(z') \rangle \, dz dz' = S_0 \int_{(k-1)\Delta z}^{k\Delta z} \int_{(k-1)\Delta z}^{k\Delta z} \delta(z-z') \, dz dz'$$
$$= S_0 \int_{(k-1)\Delta z}^{k\Delta z} dz = S_0 \Delta z.$$

These results are used throughout Chapter 3.

Thus, the expected responses are [1]

$$\mathcal{I}_0(z) = \langle I_0(z) \rangle = e^{-\Gamma_0 z} e^{-\frac{S_0}{2} z} \langle I_0(0) \rangle,$$
$$\mathcal{I}_1(z) = \langle I_1(z) \rangle = e^{-\Gamma_1 z} e^{-\frac{S_0}{2} z} \langle I_1(0) \rangle,$$
(3.30)

where $I_0(0)$ and $I_1(0)$ are, respectively, the sinusoidal signal and spurious mode inputs. By Equation (3.6), the expected normalized responses are

$$\mathcal{G}_0(z) = \langle G_0(z) \rangle = e^{-\frac{S_0}{2} z} \langle G_0(0) \rangle,$$
$$\mathcal{G}_1(z) = \langle G_1(z) \rangle = e^{-\frac{S_0}{2} z} \langle G_1(0) \rangle.$$
(3.31)

The expected complex wave amplitudes in the two-mode case decay exponentially, and are uncoupled. The larger the white coupling spectral density S_0, the more rapid the decay. For unit sinusoidal signal input and zero spurious mode input,

$$I_0(0) = G_0(0) = 1, \qquad I_1(0) = G_1(0) = 0,$$
(3.32)

the expected responses are

$$\mathcal{I}_0(z) = \langle I_0(z) \rangle = e^{-\Gamma_0 z} e^{-\frac{S_0}{2} z}, \qquad \mathcal{I}_1(z) = \langle I_1(z) \rangle = 0.$$
$$\mathcal{G}_0(z) = \langle G_0(z) \rangle = e^{-\frac{S_0}{2} z}, \qquad \mathcal{G}_1(z) = \langle G_1(z) \rangle = 0.$$
(3.33)

Here the spurious mode has expected complex amplitude identically zero for all z. Of course, zero expected value tells nothing about higher order statistics; a complex vector with unit magnitude and random uniform phase has zero average value.

The multi-mode case exhibits qualitative differences. From Equations (2.63), (2.58), and (2.45),

$$I(k\Delta z) = e^{jc_k \mathbf{C}} \cdot e^{-\Gamma \Delta z} \cdot I[(k-1)\Delta z].$$
(3.34)

$$\mathbf{C} = \begin{bmatrix} 0 & C_{01} & C_{02} & \cdots \\ C_{01} & 0 & C_{12} & \cdots \\ C_{02} & C_{12} & 0 & \cdots \\ \vdots & \vdots & \vdots & \ddots \end{bmatrix}, \qquad \mathbf{C} = \mathbf{C}^\mathrm{T} = \mathbf{C}^*.$$
(3.35)

$$\boldsymbol{\Gamma} = \begin{bmatrix} \Gamma_0 & 0 & 0 & \cdots \\ 0 & \Gamma_1 & 0 & \cdots \\ 0 & 0 & \Gamma_2 & \cdots \\ \vdots & \vdots & \vdots & \ddots \end{bmatrix}. \tag{3.36}$$

Extending the two-mode analysis above,

$$\mathcal{I}(k\Delta z) = \langle e^{jc_k \mathbf{C}} \cdot e^{-\boldsymbol{\Gamma}\Delta z} \rangle \cdot \mathcal{I}[(k-1)]\Delta z, \tag{3.37}$$

where from Equation 3.26

$$\begin{aligned} \mathcal{I}^{\mathrm{T}}(z) &= \begin{bmatrix} \mathcal{I}_0(z) & \mathcal{I}_1(z) & \cdots \end{bmatrix} = \langle I^{\mathrm{T}}(z) \rangle \\ &= \begin{bmatrix} \langle I_0(z) \rangle & \langle I_1(z) \rangle & \cdots \end{bmatrix}. \end{aligned} \tag{3.38}$$

For small Δz,

$$\begin{aligned} e^{jc_k \mathbf{C}} &\approx \mathcal{I} + jc_k \mathbf{C} - \frac{1}{2} c_k^2 \mathbf{C}^2, \\ e^{-\boldsymbol{\Gamma}\Delta z} &\approx \mathcal{I} - \boldsymbol{\Gamma}\Delta z, \end{aligned} \tag{3.39}$$

where $\mathcal{I}$ is the unit matrix. The first factor on the right-hand side of Equation (3.37) becomes

$$\langle e^{jc_k \mathbf{C}} \cdot e^{-\boldsymbol{\Gamma}\Delta z} \rangle \approx \mathcal{I} - \boldsymbol{\Gamma}\Delta z - \frac{1}{2}\langle c_k^2 \rangle \mathbf{C}^2 = \mathcal{I} - \boldsymbol{\Gamma}\Delta z - \frac{1}{2}S_0 \mathbf{C}^2 \Delta z. \tag{3.40}$$

Substituting Equation (3.40) into Equation (3.37) and taking the limit as $\Delta z \to 0$, the multi-mode extension of Equation (3.29) becomes

$$\mathcal{I}'(z) = \left(-\boldsymbol{\Gamma} - \frac{1}{2}S_0 \mathbf{C}^2\right) \cdot \mathcal{I}(z), \quad \mathcal{I}(z) = \langle I(z) \rangle. \tag{3.41}$$

The substitution

$$\mathbf{C} = \begin{bmatrix} 0 & 1 \\ 1 & 0 \end{bmatrix}, \quad \mathbf{C}^2 = \mathcal{I}, \tag{3.42}$$

reduces Equation (3.41) to Equation (3.29) for the two-mode case.

It is instructive to write explicit results for the three-mode case:

$$\mathbf{C} = \begin{bmatrix} 0 & C_{01} & C_{02} \\ C_{01} & 0 & C_{12} \\ C_{02} & C_{12} & 0 \end{bmatrix},$$

$$\mathbf{C}^2 = \begin{bmatrix} C_{01}^2 + C_{02}^2 & C_{02}C_{12} & C_{01}C_{12} \\ C_{02}C_{12} & C_{01}^2 + C_{12}^2 & C_{01}C_{02} \\ C_{01}C_{12} & C_{01}C_{02} & C_{02}^2 + C_{12}^2 \end{bmatrix}.$$
(3.43)

The differential equations for the expected complex mode amplitudes are

$$\mathcal{I}_0'(z) = -\left\{\Gamma_0 + \frac{1}{2}S_0(C_{01}^2 + C_{02}^2)\right\}\mathcal{I}_0(z)$$

$$- \frac{1}{2}S_0 C_{02} C_{12} \cdot \mathcal{I}_1(z) - \frac{1}{2}S_0 C_{01} C_{12} \cdot \mathcal{I}_2(z). \quad (3.44)$$

$$\mathcal{I}_1'(z) = -\frac{1}{2}S_0 C_{02} C_{12} \cdot \mathcal{I}_0(z) - \left\{\Gamma_1 + \frac{1}{2}S_0(C_{01}^2 + C_{12}^2)\right\}\mathcal{I}_1(z)$$

$$- \frac{1}{2}S_0 C_{01} C_{02} \cdot \mathcal{I}_2(z). \quad (3.45)$$

$$\mathcal{I}_2'(z) = -\frac{1}{2}S_0 C_{01} C_{12} \cdot \mathcal{I}_0(z) - \frac{1}{2}S_0 C_{01} C_{02} \cdot \mathcal{I}_1(z)$$

$$- \left\{\Gamma_2 + \frac{1}{2}S_0(C_{02}^2 + C_{12}^2)\right\}\mathcal{I}_2(z). \quad (3.46)$$

The expected complex mode amplitudes are no longer uncoupled. Even for zero spurious mode inputs

$$I_0(0) = 1, \quad I_1(0) = I_2(0) = 0, \quad (3.47)$$

the spurious mode averages $\mathcal{I}_1(z) = \langle I_1(z) \rangle$ and $\mathcal{I}_2(z) = \langle I_2(z) \rangle$ depart from zero for a coherent input signal.

We use coherent inputs to the coupled line equations to study transfer function statistics, here the expected transfer functions. For a coherent sinusoidal input with unit power to mode i we set $I_i(0) = 1$, a deterministic quantity representing a complex wave $e^{j\omega t}$. Sources such as light-emitting diodes (LEDs) produce incoherent or partially coherent inputs to multi-mode optical fibers. We model an incoherent sinusoidal input of unit power with constant amplitude and random phase as $I_i(0) = e^{j\theta}$, with θ a random variable uniformly distributed from $0 < \theta < 2\pi$, representing the complex wave $e^{j\omega t} e^{j\theta}$; here the expected input is zero, $\langle I_0(0) \rangle = 0$. The present results, Equations (3.29), (3.41), or (3.44)–(3.46), show that all expected mode amplitudes are zero for incoherent inputs; for coherent or partially coherent inputs the expected mode amplitudes depart from zero.

The present results simplify in the nondegenerate case, where no two modes have the same propagation constants. We show in Section 4.6.1 that for large $\Delta\beta_{ij} = \beta_i - \beta_j$ the coupling between expected mode amplitudes approaches zero, as in the two-mode case.

The expected complex transfer functions give little information about the transmission behavior of guides with random coupling. Exact results for higher order statistics—average powers, power fluctuations, transfer function vs. frequency statistics—are calculated below using Kronecker products.

3.4. COUPLED POWER EQUATIONS

For two modes with constant propagation parameters, take the Kronecker product of Equation (3.24) with its complex conjugate, to yield by Equation (C.7)

$$\begin{bmatrix} |I_0(k\Delta z)|^2 \\ I_0(k\Delta z)I_1^*(k\Delta z) \\ I_1(k\Delta z)I_0^*(k\Delta z) \\ |I_1(k\Delta z)|^2 \end{bmatrix} = (\mathbf{M} \otimes \mathbf{M}^*) \cdot \begin{bmatrix} |I_0[(k-1)\Delta z]|^2 \\ I_0[(k-1)\Delta z]I_1^*[(k-1)\Delta z] \\ I_1[(k-1)\Delta z]I_0^*[(k-1)\Delta z] \\ |I_1[(k-1)\Delta z]|^2 \end{bmatrix},$$

(3.48)

3.4. COUPLED POWER EQUATIONS

where **M** is the 2 × 2 matrix on the right-hand side of Equation (3.24), and the elements of $\mathbf{M} \otimes \mathbf{M}^*$ are given as follows:

$$
\begin{aligned}
(\mathbf{M} \otimes \mathbf{M}^*)_{11} &= e^{-\Gamma_0 \Delta z} \cos c_k \cdot e^{-\Gamma_0^* \Delta z} \cos c_k \\
(\mathbf{M} \otimes \mathbf{M}^*)_{12} &= -e^{-\Gamma_0 \Delta z} \cos c_k \cdot e^{-\Gamma_1^* \Delta z} j \sin c_k \\
(\mathbf{M} \otimes \mathbf{M}^*)_{13} &= e^{-\Gamma_1 \Delta z} j \sin c_k \cdot e^{-\Gamma_0^* \Delta z} \cos c_k \\
(\mathbf{M} \otimes \mathbf{M}^*)_{14} &= -e^{-\Gamma_1 \Delta z} j \sin c_k \cdot e^{-\Gamma_1^* \Delta z} j \sin c_k \\
(\mathbf{M} \otimes \mathbf{M}^*)_{21} &= -e^{-\Gamma_0 \Delta z} \cos c_k \cdot e^{-\Gamma_0^* \Delta z} j \sin c_k \\
(\mathbf{M} \otimes \mathbf{M}^*)_{22} &= e^{-\Gamma_0 \Delta z} \cos c_k \cdot e^{-\Gamma_1^* \Delta z} \cos c_k \\
(\mathbf{M} \otimes \mathbf{M}^*)_{23} &= -e^{-\Gamma_1 \Delta z} j \sin c_k \cdot e^{-\Gamma_0^* \Delta z} j \sin c_k \\
(\mathbf{M} \otimes \mathbf{M}^*)_{24} &= e^{-\Gamma_1 \Delta z} j \sin c_k \cdot e^{-\Gamma_1^* \Delta z} \cos c_k \\
(\mathbf{M} \otimes \mathbf{M}^*)_{31} &= e^{-\Gamma_0 \Delta z} j \sin c_k \cdot e^{-\Gamma_0^* \Delta z} \cos c_k \\
(\mathbf{M} \otimes \mathbf{M}^*)_{32} &= -e^{-\Gamma_0 \Delta z} j \sin c_k \cdot e^{-\Gamma_1^* \Delta z} j \sin c_k \\
(\mathbf{M} \otimes \mathbf{M}^*)_{33} &= e^{-\Gamma_1 \Delta z} \cos c_k \cdot e^{-\Gamma_0^* \Delta z} \cos c_k \\
(\mathbf{M} \otimes \mathbf{M}^*)_{34} &= -e^{-\Gamma_1 \Delta z} \cos c_k \cdot e^{-\Gamma_1^* \Delta z} j \sin c_k \\
(\mathbf{M} \otimes \mathbf{M}^*)_{41} &= -e^{-\Gamma_0 \Delta z} j \sin c_k \cdot e^{-\Gamma_0^* \Delta z} j \sin c_k \\
(\mathbf{M} \otimes \mathbf{M}^*)_{42} &= e^{-\Gamma_0 \Delta z} j \sin c_k \cdot e^{-\Gamma_1^* \Delta z} \cos c_k \\
(\mathbf{M} \otimes \mathbf{M}^*)_{43} &= -e^{-\Gamma_1 \Delta z} \cos c_k \cdot e^{-\Gamma_0^* \Delta z} j \sin c_k \\
(\mathbf{M} \otimes \mathbf{M}^*)_{44} &= e^{-\Gamma_1 \Delta z} \cos c_k \cdot e^{-\Gamma_1^* \Delta z} \cos c_k.
\end{aligned}
\tag{3.49}
$$

As before, the restrictions of Equation (3.25) must be satisfied. For white coupling, different sections are independent for any Δz, and the expected value of Equation (3.48) may be identified with Equation (D.17). Define the average powers and cross-powers as follows:[2]

$$
\begin{aligned}
\mathcal{P}_i(z) &= \langle P_i(z) \rangle = \langle |I_i(z)|^2 \rangle; \\
\mathcal{P}_{ij}(z) &= \mathcal{P}_{ji}^*(z) = \langle P_{ij}(z) \rangle, \quad P_{ij}(z) = I_i(z) I_j^*(z).
\end{aligned}
\tag{3.50}
$$

[2] $P_i(z)$ is defined in Equations (2.3) and (2.46).

Taking expected values in Equations (3.48) and (3.49), substituting Equation (3.50), evaluating

$$\langle \cos^2 c_k \rangle \approx 1 - \langle c_k^2 \rangle = 1 - S_0 \Delta z,$$
$$\langle \sin^2 c_k \rangle \approx \langle c_k^2 \rangle = S_0 \Delta z, \qquad (3.51)$$
$$\langle \sin c_k \cos c_k \rangle \approx \langle c_k \rangle = 0,$$

as in Equation (3.28), and taking the limit as $\Delta z \to 0$,

$$\begin{bmatrix} \mathcal{P}'_0(z) \\ \mathcal{P}'_{01}(z) \\ \mathcal{P}'_{10}(z) \\ \mathcal{P}'_1(z) \end{bmatrix} = \begin{bmatrix} -(2\alpha_0 + S_0) & 0 & 0 & S_0 \\ 0 & -(\Gamma_0 + \Gamma_1^* + S_0) & S_0 & 0 \\ 0 & S_0 & -(\Gamma_0^* + \Gamma_1 + S_0) & 0 \\ S_0 & 0 & 0 & -(2\alpha_1 + S_0) \end{bmatrix} \cdot \begin{bmatrix} \mathcal{P}_0(z) \\ \mathcal{P}_{01}(z) \\ \mathcal{P}_{10}(z) \\ \mathcal{P}_1(z) \end{bmatrix}. \qquad (3.52)$$

Equation (3.52) separates into two sets of differential equations. The average mode powers are [1, 2]

$$\mathcal{P}'_0(z) = -(2\alpha_0 + S_0)\mathcal{P}_0(z) + S_0 \mathcal{P}_1(z).$$
$$\mathcal{P}'_1(z) = S_0 \mathcal{P}_0(z) - (2\alpha_1 + S_0)\mathcal{P}_1(z). \qquad (3.53)$$

The cross-powers are [1]

$$\mathcal{P}'_{01}(z) = -(\Gamma_0 + \Gamma_1^* + S_0)\mathcal{P}_{01}(z) + S_0 \mathcal{P}_{10}(z).$$
$$\mathcal{P}'_{10}(z) = S_0 \mathcal{P}_{01}(z) - (\Gamma_0^* + \Gamma_1 + S_0)\mathcal{P}_{10}(z). \qquad (3.54)$$

The two relations of Equation (3.54) are complex conjugates of each other.

3.4. COUPLED POWER EQUATIONS

The powers and cross-powers are decoupled only in the two-mode case. In general, the expected value of the Kronecker product of Equation (3.34) with its complex conjugate yields

$$\langle \boldsymbol{I}(k\Delta z) \otimes \boldsymbol{I}^*(k\Delta z) \rangle = \langle \mathbf{M} \otimes \mathbf{M}^* \rangle \cdot \langle \boldsymbol{I}[(k-1)\Delta z] \otimes \boldsymbol{I}^*[(k-1)\Delta z] \rangle, \tag{3.55}$$

where

$$\mathbf{M} = e^{jc_k \mathbf{C}} \cdot e^{-\Gamma \Delta z}. \tag{3.56}$$

For N modes, define the Kronecker product of Equation 2.43 with its complex conjugate as

$$\boldsymbol{\mathcal{P}}^T(z) = \langle \boldsymbol{P}^T(z) \rangle, \quad \boldsymbol{P}^T(z) = \boldsymbol{I}^T(z) \otimes \boldsymbol{I}^{T*}(z),$$
$$\boldsymbol{\mathcal{P}}^T(z) = [\mathcal{P}_0 \, \mathcal{P}_{01} \cdots \mathcal{P}_{0,N-1} \quad \mathcal{P}_{10} \, \mathcal{P}_1 \cdots \mathcal{P}_{1,N-1} \quad \cdots$$
$$\cdots \quad \mathcal{P}_{N-1,0} \, \mathcal{P}_{N-1,1} \cdots \mathcal{P}_{N-1}], \tag{3.57}$$

where the $\mathcal{P}$'s represent the average powers and cross-powers defined in Equation (3.50). Each of these elements is a function of z, i.e., $\mathcal{P}(z)$; this dependence has been suppressed for compactness. Then, Equation (3.55) becomes

$$\boldsymbol{\mathcal{P}}(k\Delta z) = \langle \mathbf{M} \otimes \mathbf{M}^* \rangle \cdot \boldsymbol{\mathcal{P}}[(k-1)\Delta z]. \tag{3.58}$$

From Equation (3.56),

$$\mathbf{M} \otimes \mathbf{M}^* = (e^{jc_k \mathbf{C}} \cdot e^{-\Gamma \Delta z}) \otimes (e^{-jc_k \mathbf{C}} \cdot e^{-\Gamma^* \Delta z})$$
$$= (e^{jc_k \mathbf{C}} \otimes e^{-jc_k \mathbf{C}}) \cdot (e^{-\Gamma \Delta z} \otimes e^{-\Gamma^* \Delta z}). \tag{3.59}$$

Substituting Equation (3.39) and taking expected values, for small Δz

$$\langle \mathbf{M} \otimes \mathbf{M}^* \rangle \approx \langle (\mathcal{I} + jc_k \mathbf{C} - \tfrac{1}{2} c_k^2 \mathbf{C}^2) \otimes (\mathcal{I} - jc_k \mathbf{C} - \tfrac{1}{2} c_k^2 \mathbf{C}^2) \rangle$$
$$\cdot (\mathcal{I} - \Gamma \Delta z) \otimes (\mathcal{I} - \Gamma^* \Delta z)$$
$$= \{\mathcal{I} \otimes \mathcal{I} + S_0 \Delta z (-\tfrac{1}{2} \mathcal{I} \otimes \mathbf{C}^2 + \mathbf{C} \otimes \mathbf{C} - \tfrac{1}{2} \mathbf{C}^2 \otimes \mathcal{I})\}$$
$$\cdot \{\mathcal{I} \otimes \mathcal{I} - \Delta z (\mathcal{I} \otimes \Gamma^* + \Gamma \otimes \mathcal{I})\}$$
$$= \mathcal{I} \otimes \mathcal{I} + \Delta z \{S_0 (-\tfrac{1}{2} \mathcal{I} \otimes \mathbf{C}^2 + \mathbf{C} \otimes \mathbf{C} - \tfrac{1}{2} \mathbf{C}^2 \otimes \mathcal{I})$$
$$- \mathcal{I} \otimes \Gamma^* - \Gamma \otimes \mathcal{I}\}. \tag{3.60}$$

Substituting Equation (3.60) into Equation (3.58) and taking the limit as $\Delta z \to 0$, the multi-mode generalization of Equations (3.52) to (3.54) is

$$\mathcal{P}'(z) = \{S_0(\mathbf{C} \otimes \mathbf{C} - \tfrac{1}{2}\mathcal{I} \otimes \mathbf{C}^2 - \tfrac{1}{2}\mathbf{C}^2 \otimes \mathcal{I}) - \mathcal{I} \otimes \mathbf{\Gamma}^* - \mathbf{\Gamma} \otimes \mathcal{I}\} \cdot \mathcal{P}(z),$$
(3.61)

where $\mathbf{C}$ and $\mathbf{\Gamma}$ are given by Equations (3.35) and (3.36) and $\mathcal{I}$ is the $N \times N$ unit matrix. Equation (3.61) reduces to the two-mode case by the substitution of Equation (3.42).

Even the two-mode case involves a disagreeable amount of hand calculation, requiring the expected values of 16 terms of a 4×4 matrix. It is evident that greater numbers of modes require the use of a computer algebra system capable of programmable symbolic and numeric algebraic, matrix, and calculus operations. The N-mode coupled power equations of Equation (3.61) comprise a set of N^2 differential equations; N of these are real, $N^2 - N$ occur in complex conjugate pairs.

The three-mode case of Equation (3.43) leads to a 9×9 matrix with 81 terms, yielding 9 differential equations relating the average powers and cross powers, each with 9 terms on the right-hand side. These have been obtained with MAPLE; the first two of these 9 equations are given below:

$$\begin{aligned}
\mathcal{P}'_0(z) = & -[S_0(C_{01}^2 + C_{02}^2) + \Gamma_0 + \Gamma_0^*]\mathcal{P}_0(z) - .5S_0 C_{02} C_{12} \mathcal{P}_{01}(z) \\
& - .5S_0 C_{01} C_{12} \mathcal{P}_{02}(z) - .5S_0 C_{02} C_{12} \mathcal{P}_{10}(z) + S_0 C_{01}^2 \mathcal{P}_1(z) \\
& + S_0 C_{01} C_{02} \mathcal{P}_{12}(z) - .5S_0 C_{01} C_{12} \mathcal{P}_{20}(z) + S_0 C_{01} C_{02} \mathcal{P}_{21}(z) \\
& + S_0 C_{02}^2 \mathcal{P}_2(z). \\
\mathcal{P}'_{01}(z) = & -.5S_0 C_{02} C_{12} \mathcal{P}_0(z) \\
& - [S_0(C_{01}^2 + .5C_{12}^2 + .5C_{02}^2) + \Gamma_0 + \Gamma_1^*]\mathcal{P}_{01}(z) \\
& - .5S_0 C_{01} C_{02} \mathcal{P}_{02}(z) + S_0 C_{01}^2 \mathcal{P}_{10}(z) - .5S_0 C_{02} C_{12} \mathcal{P}_1(z) \\
& + S_0 C_{01} C_{12} \mathcal{P}_{12}(z) + S_0 C_{01} C_{02} \mathcal{P}_{20}(z) - .5S_0 C_{01} C_{12} \mathcal{P}_{21}(z) \\
& + S_0 C_{02} C_{12} \mathcal{P}_2(z).
\end{aligned}$$

$$\dots\dots\dots\dots\dots\dots\dots\dots\dots\dots\dots\dots\dots\dots\dots\dots$$
(3.62)

It is unlikely that anyone will wish to deal with even three modes by hand. Subsequent examples given in Chapters 4, 6, 7, and 8 demonstrate that MAPLE is capable of treating significant multimode problems.

The coupled power equations of this section apply equally well to coherent, incoherent, or partially coherent inputs; their solutions yield the mode powers as a function of distance z along the guide in terms of the input mode powers, whatever the state of coherence at the input. For coherent or partially coherent inputs, the modes will be correlated and will contain coherent components given by the results of Section 3.3; but the total mode powers and cross-powers depend only on their input values.

Strict decoupling of the average powers and cross-powers occurs only in the two-mode case. However, if all propagation constants are distinct (i.e., no degenerate modes), the powers and the cross-powers are approximately decoupled. In this approximation, the average mode powers are described by N differential equations [2], rather than the N^2 differential equations of Equation (3.61). Further discussion is given in Section 4.6.2.

3.5. POWER FLUCTUATIONS

Higher order statistics can be readily obtained in a similar way. As an example we calculate the power fluctuations for white coupling. Define

$$P_{ijk\ell}(z) = I_i(z)I_j^*(z)I_k(z)I_\ell^*(z). \qquad (3.63)$$

The average of this quantity is the fourth-order moment

$$\mathcal{P}_{ijk\ell}(z) = \langle P_{ijk\ell}(z) \rangle = \langle I_i(z)I_j^*(z)I_k(z)I_\ell^*(z) \rangle. \qquad (3.64)$$

Powers and cross-powers previously defined in Equations (3.50), (2.3), and (2.46) occur in some of these quantities. Of particular present interest

$$P_{iiii}(z) = P_i^2(z) \qquad (3.65)$$

is the square of the power in the ith mode,

$$\mathcal{P}_{iiii}(z) = \langle P_{iiii}(z) \rangle = \langle P_i^2(z) \rangle \qquad (3.66)$$

is the second moment of the ith mode power, and

$$\begin{aligned}\mathcal{P}_{iijj}(z) &= \langle P_i(z)P_j(z)\rangle, \quad \mathcal{P}_{ijij}(z) = \langle P_{ij}^2(z)\rangle, \\ \mathcal{P}_{ijji}(z) &= \langle |P_{ij}(z)|^2 \rangle = \langle P_i(z)P_j(z)\rangle, \qquad i \neq j.\end{aligned} \qquad (3.67)$$

Define the mean-square power fluctuation in the ith mode as

$$\langle [\Delta P_i(z)]^2 \rangle = \langle P_i^2(z)\rangle - \langle P_i(z)\rangle^2 = \mathcal{P}_{iiii}(z) - \mathcal{P}_i^2(z). \qquad (3.68)$$

$\mathcal{P}_i(z)$ was found in Section 3.4; we determine $\mathcal{P}_{iiii}(z)$ in this section, and through Equation (3.68), the power fluctuation. From Equation (3.34),

$$\begin{aligned}\langle I(k\Delta z) \otimes I^*(k\Delta z) &\otimes I(k\Delta z) \otimes I^*(k\Delta z) \rangle \\ = \langle e^{jc_k C} &\otimes e^{-jc_k C} \otimes e^{jc_k C} \otimes e^{-jc_k C}\rangle \\ \cdot (e^{-\Gamma \Delta z} &\otimes e^{-\Gamma^* \Delta z} \otimes e^{-\Gamma \Delta z} \otimes e^{-\Gamma^* \Delta z}) \\ \cdot \langle I[(k-1)\Delta z] &\otimes I^*[(k-1)\Delta z] \\ \otimes I[(k-1)\Delta z] &\otimes I^*[(k-1)\Delta z]\rangle.\end{aligned} \qquad (3.69)$$

Define

$$\begin{aligned}\mathcal{P}_4^{\mathrm{T}}(z) &= \langle I^{\mathrm{T}}(z) \otimes I^{\mathrm{T}*}(z) \otimes I^{\mathrm{T}}(z) \otimes I^{\mathrm{T}*}\rangle \\ &= [\mathcal{P}_{0000}(z) \quad \mathcal{P}_{0001}(z) \cdots \mathcal{P}_{ijk\ell}(z) \cdots \mathcal{P}_{N-1,N-1,N-1,N-1}(z)],\end{aligned} \qquad (3.70)$$

where the subscripts, $\mathrm{mod}(N)$, are arranged in numerical order. Substituting Equation (3.70) into Equation (3.69),

$$\begin{aligned}\mathcal{P}_4(k\Delta z) = \langle e^{jc_k C} &\otimes e^{-jc_k C} \otimes e^{jc_k C} \otimes e^{-jc_k C}\rangle \\ \cdot (e^{-\Gamma \Delta z} &\otimes e^{-\Gamma^* \Delta z} \otimes e^{-\Gamma \Delta z} \otimes e^{-\Gamma^* \Delta z}) \\ \cdot \mathcal{P}_4[(k-1)&\Delta z].\end{aligned} \qquad (3.71)$$

3.5. POWER FLUCTUATIONS

Substituting Equation (3.39) and taking expected values, the limit as $\Delta z \to 0$ yields the following coupled differential equations for the fourth moments, for N modes:

$$\mathcal{P}_4'(z) = \Big[S_0 \Big\{ \mathcal{I} \otimes \mathcal{I} \otimes \mathbf{C} \otimes \mathbf{C} - \mathcal{I} \otimes \mathbf{C} \otimes \mathcal{I} \otimes \mathbf{C} + \mathcal{I} \otimes \mathbf{C} \otimes \mathbf{C} \otimes \mathcal{I}$$

$$+ \mathbf{C} \otimes \mathcal{I} \otimes \mathcal{I} \otimes \mathbf{C} - \mathbf{C} \otimes \mathcal{I} \otimes \mathbf{C} \otimes \mathcal{I} + \mathbf{C} \otimes \mathbf{C} \otimes \mathcal{I} \otimes \mathcal{I}$$

$$- \frac{1}{2}(\mathcal{I} \otimes \mathcal{I} \otimes \mathcal{I} \otimes \mathbf{C}^2 + \mathcal{I} \otimes \mathcal{I} \otimes \mathbf{C}^2 \otimes \mathcal{I}$$

$$+ \mathcal{I} \otimes \mathbf{C}^2 \otimes \mathcal{I} \otimes \mathcal{I} + \mathbf{C}^2 \otimes \mathcal{I} \otimes \mathcal{I} \otimes \mathcal{I}) \Big\}$$

$$- \mathcal{I} \otimes \mathcal{I} \otimes \mathcal{I} \otimes \mathbf{\Gamma}^* - \mathcal{I} \otimes \mathcal{I} \otimes \mathbf{\Gamma} \otimes \mathcal{I}$$

$$- \mathcal{I} \otimes \mathbf{\Gamma}^* \otimes \mathcal{I} \otimes \mathcal{I} - \mathbf{\Gamma} \otimes \mathcal{I} \otimes \mathcal{I} \otimes \mathcal{I} \Big] \cdot \mathcal{P}_4(z), \quad (3.72)$$

where $\mathcal{I}$ is the $N \times N$ unit matrix, and $\mathbf{C}$ and $\mathbf{\Gamma}$ are given by Equations (3.35) and (3.36).

Equation (3.72) is exact; it represents N^4 first-order differential equations for the fourth moments, each with N^4 right-hand-side terms. However, eliminating redundant equations yields a significant reduction in the number of equations. By elementary combinatorics, for $N \geq 4$ eliminating identical equations reduces the number from N^4 to $N^4/4 + N^3/2 + N^2/4$. Further reductions occur in special cases, illustrated below for $N = 2$ modes.

To determine the power fluctuations in the two-mode case we substitute Equation (3.42) into the right-hand side of Equation (3.72), obtaining 16 simultaneous first-order differential equations. The resulting matrix has been evaluated by MAPLE; significant simplifications occur. The 16 differential equations separate into two sets of 8 equations each, one set containing only the functions of Equations (3.66) and (3.67) with $i, j = 0, 1$, the other containing mixed moments, i.e., quantities of Equation (3.64) with more than two distinct subscripts. The first set, of current interest, contains 5 independent equations. Define

$$\mathbf{Q}^T(z) = [\mathcal{P}_{0000}(z) \quad \mathcal{P}_{0011}(z) \quad \mathcal{P}_{0101}(z) \quad \mathcal{P}_{1010}(z) \quad \mathcal{P}_{1111}(z)],$$
$$(3.73)$$

where from Equations (3.66) and (3.67)

$$\begin{aligned}
\mathcal{P}_{0000}(z) &= \langle I_0(z)I_0^*(z)I_0(z)I_0^*(z)\rangle = \langle P_0^2(z)\rangle \\
\mathcal{P}_{0011}(z) &= \langle I_0(z)I_0^*(z)I_1(z)I_1^*(z)\rangle = \langle P_0(z)P_1(z)\rangle \\
\mathcal{P}_{0101}(z) &= \langle I_0(z)I_1^*(z)I_0(z)I_1^*(z)\rangle = \langle P_{01}^2(z)\rangle \\
\mathcal{P}_{1010}(z) &= \langle I_1(z)I_0^*(z)I_1(z)I_0^*(z)\rangle = \langle P_{10}^2(z)\rangle = \langle P_{01}^2(z)\rangle^* \\
\mathcal{P}_{1111}(z) &= \langle I_1(z)I_1^*(z)I_1(z)I_1^*(z)\rangle = \langle P_1^2(z)\rangle.
\end{aligned} \quad (3.74)$$

Then the fourth moments of interest are given by the differential equations

$$\mathbf{Q}'(z) = \{S_0 \mathbf{N}_1 - \mathbf{N}_2\} \cdot \mathbf{Q}(z), \quad (3.75)$$

where S_0, the coupling spectral density, is given in Equations (3.1) and (3.2), and the matrices $\mathbf{N}_1$ and $\mathbf{N}_2$ are given as follows:

$$\mathbf{N}_1 = \begin{bmatrix} -2 & 4 & -1 & -1 & 0 \\ 1 & -4 & 1 & 1 & 1 \\ -1 & 4 & -2 & 0 & -1 \\ -1 & 4 & 0 & -2 & -1 \\ 0 & 4 & -1 & -1 & -2 \end{bmatrix}, \quad (3.76)$$

$$\mathbf{N}_2 = \begin{bmatrix} -4\alpha_0 & 0 & 0 & 0 & 0 \\ 0 & -2\alpha_+ & 0 & 0 & 0 \\ 0 & 0 & -2(\alpha_+ + j\Delta\beta) & 0 & 0 \\ 0 & 0 & 0 & -2(\alpha_+ - j\Delta\beta) & 0 \\ 0 & 0 & 0 & 0 & -4\alpha_1 \end{bmatrix}, \quad (3.77)$$

where

$$\alpha_+ = \alpha_0 + \alpha_1, \qquad \Delta\beta = \beta_0 - \beta_1. \quad (3.78)$$

The present calculations for power fluctuations with white coupling are exact. Cross-powers and higher mixed moments are included without approximations or assumptions of any kind. For two modes the calculation for power fluctuations includes the cross-correlation of the mode powers $\langle P_0(z)P_1(z)\rangle$ [2], as well as the square of the cross-power $\langle P_{01}^2(z)\rangle$. Finally, further simplification occurs in the non-degenerate case, described in Section 4.6.3.

3.6. TRANSFER FUNCTION STATISTICS

We have so far considered only single-frequency transmission statistics, in Sections 3.3, 3.4, and 3.5. Multi-frequency statistics are necessary in order to describe signal distortion. We therefore consider the second-order gain-frequency statistics for a two-mode guide. We adopt the notation of Section 3.2, with initial conditions given by Equation (3.14); $G_0(z)$, the normalized signal transfer function, is a function of $\Delta\beta$. It proves convenient to define the following new quantity:

$$G_{01}(z) = e^{\Gamma_0 z} I_1(z), \qquad G_{01}(0) = 0. \tag{3.79}$$

$G_{01}(z)$ is the normalized signal-spurious mode transfer function, also a function of $\Delta\beta$. Then, Equation (3.24) becomes

$$\begin{bmatrix} G_0(k\Delta z) \\ G_{01}(k\Delta z) \end{bmatrix} = \begin{bmatrix} \cos c_k & e^{\Delta\Gamma\cdot\Delta z} j \sin c_k \\ j \sin c_k & e^{\Delta\Gamma\cdot\Delta z} \cos c_k \end{bmatrix} \cdot \begin{bmatrix} G_0[(k-1)\Delta z] \\ G_{01}[(k-1)\Delta z] \end{bmatrix}. \tag{3.80}$$

We rewrite this relation as

$$\begin{bmatrix} G_0(\Delta\beta) \\ G_{01}(\Delta\beta) \end{bmatrix}_{k\Delta z} = \begin{bmatrix} \cos c_k & e^{\Delta\Gamma\cdot\Delta z} j \sin c_k \\ j \sin c_k & e^{\Delta\Gamma\cdot\Delta z} \cos c_k \end{bmatrix} \cdot \begin{bmatrix} G_0(\Delta\beta) \\ G_{01}(\Delta\beta) \end{bmatrix}_{(k-1)\Delta z}, \tag{3.81}$$

in which the z dependence of the column vectors is indicated by subscripts, and the $\Delta\alpha$ dependence is implicit.

We assume the frequency dependence occurs through $\Delta\beta$, and neglect the frequency dependence of c_k and $\Delta\alpha$, following the discussion in Section 3.2. Set $\Delta\beta \to \Delta\beta + \sigma$ in Equation (3.81):

$$\begin{bmatrix} G_0(\Delta\beta + \sigma) \\ G_{01}(\Delta\beta + \sigma) \end{bmatrix}_{k\Delta z} = \begin{bmatrix} \cos c_k & \pm e^{(\Delta\Gamma + j\sigma)\Delta z} j \sin c_k \\ \pm j \sin c_k & e^{(\Delta\Gamma + j\sigma)\Delta z} \cos c_k \end{bmatrix}$$

$$\cdot \begin{bmatrix} G_0(\Delta\beta + \sigma) \\ G_{01}(\Delta\beta + \sigma) \end{bmatrix}_{(k-1)\Delta z}, \quad (3.82)$$

where the $+$ signs are taken when $\Delta\beta$ and $\Delta\beta + \sigma$ have the same sign (i.e., both frequencies are positive, as in the usual case, or both are negative), the $-$ signs when they have opposite signs, in accord with Equations (3.15) and (3.18).

The analysis follows that of Section 3.4:

1. Take the Kronecker product of Equation (3.82) with the complex conjugate of Equation (3.81).
2. Take the expected value of the result.
3. Substitute the following definitions for the resulting column vectors:

$$\begin{bmatrix} \mathcal{R}_0(z) \\ \mathcal{R}_{01}(z) \\ \mathcal{R}_{10}(z) \\ \mathcal{R}_1(z) \end{bmatrix}_\sigma = \begin{bmatrix} \langle G_0(\Delta\beta + \sigma)G_0^*(\Delta\beta)\rangle \\ \langle G_0(\Delta\beta + \sigma)G_{01}^*(\Delta\beta)\rangle \\ \langle G_{01}(\Delta\beta + \sigma)G_0^*(\Delta\beta)\rangle \\ \langle G_{01}(\Delta\beta + \sigma)G_{01}^*(\Delta\beta)\rangle \end{bmatrix}_z. \quad (3.83)$$

The subscript σ on the left-hand side indicates the dependence of the $\mathcal{R}$'s on the normalized frequency interval σ; the subscript z on the right-hand side indicates the z dependence of the G's, as before.

4. Substitute Equation (3.51).
5. Take the limit as $\Delta z \to 0$.

3.6. TRANSFER FUNCTION STATISTICS

The final results are as follows:

$$\begin{bmatrix} \mathcal{R}'_0(z) \\ \mathcal{R}'_{01}(z) \\ \mathcal{R}'_{10}(z) \\ \mathcal{R}'_1(z) \end{bmatrix} = \begin{bmatrix} -S_0 & 0 & 0 & \pm S_0 \\ 0 & \Delta\Gamma^* - S_0 & \pm S_0 & 0 \\ 0 & \pm S_0 & \Delta\Gamma + j\sigma - S_0 & 0 \\ \pm S_0 & 0 & 0 & 2\Delta\alpha + j\sigma - S_0 \end{bmatrix} \cdot \begin{bmatrix} \mathcal{R}_0(z) \\ \mathcal{R}_{01}(z) \\ \mathcal{R}_{10}(z) \\ \mathcal{R}_1(z) \end{bmatrix}. \quad (3.84)$$

The initial conditions are obtained from Equation (3.14) and (3.79):

$$\mathcal{R}_0(\sigma)_{z=0} = 1, \qquad \mathcal{R}_1(\sigma)_{z=0} = 0. \quad (3.85)$$

$$\mathcal{R}_{01}(\sigma)_{z=0} = \mathcal{R}_{10}(\sigma)_{z=0} = 0. \quad (3.86)$$

The four equations again decouple into two sets of two equations, in the two-mode case.

For the cross-moments, Equations (3.84) and (3.86) yield

$$\mathcal{R}_{01}(\sigma)_z = \mathcal{R}_{10}(\sigma)_z = 0, \qquad \text{all } z. \quad (3.87)$$

Signal and spurious modes remain uncorrelated for all z, for zero spurious mode input.

The transfer function covariances are defined in Equation (3.83) as

$$\begin{aligned} \mathcal{R}_0(\sigma) &= \langle G_0(\Delta\beta + \sigma)G_0^*(\Delta\beta)\rangle, \\ \mathcal{R}_1(\sigma) &= \langle G_{01}(\Delta\beta + \sigma)G_{01}^*(\Delta\beta)\rangle. \end{aligned} \quad (3.88)$$

They are obtained from Equation (3.84) as solutions of the following equations, with initial conditions given by Equation (3.85) [1]:

$$\begin{aligned} \mathcal{R}'_0(z) &= -S_0\mathcal{R}_0(z) \pm S_0\mathcal{R}_1(z), \\ \mathcal{R}'_1(z) &= \pm S_0\mathcal{R}_0(z) + (2\Delta\alpha + j\sigma - S_0)\mathcal{R}_1(z). \end{aligned} \quad (3.89)$$

The solutions yield exact expressions for the normalized signal and signal-spurious mode transfer function covariances, for white coupling, as follows:

$$\mathcal{R}_0(\sigma) = \exp[-S_0 z(1 - \Xi)]$$
$$\cdot \left[\cosh(S_0 z\sqrt{1+\Xi^2}) - \Xi \frac{\sinh(S_0 z\sqrt{1+\Xi^2})}{\sqrt{1+\Xi^2}}\right]. \quad (3.90)$$

$$\mathcal{R}_1(\sigma) = \pm \exp[-S_0 z(1 - \Xi)] \frac{\sinh(S_0 z\sqrt{1+\Xi^2})}{\sqrt{1+\Xi^2}}. \quad (3.91)$$

$$\Xi = \frac{\Delta\alpha + j\sigma/2}{S_0}. \quad (3.92)$$

Finally, from Equation (3.33)

$$\langle G_0(\Delta\beta) \rangle = e^{-\frac{S_0}{2}z}, \qquad \langle G_{01}(\Delta\beta) \rangle = 0. \quad (3.93)$$

The transfer functions are wide-sense stationary functions of normalized frequency $\Delta\beta$.

The multi-mode case is treated similarly. We assume as in Section 3.2 that the frequency dependence enters only through the phase constants β_i. In contrast, for the two-mode case only the single parameter $\Delta\beta$ varies with frequency. We append subscripts x to elements of Equation (3.34) to indicate evaluation at frequency f_x:

$$I_x(k\Delta z) = e^{jc_k C} \cdot e^{-\Gamma_x \Delta z} \cdot I_x[(k-1)\Delta z], \quad (3.94)$$

with C and Γ given by Equations (3.35) and (3.36). Define the transfer function co- and cross-variances as

$$\mathcal{R}_i(z)_{xy} = \langle I_i(z)_x I_i^*(z)_y \rangle,$$
$$\mathcal{R}_{ij}(z)_{xy} = \langle I_i(z)_x I_j^*(z)_y \rangle = \mathcal{R}_{ji}^*(z)_{yx}, \quad i \neq j. \quad (3.95)$$

Define the covariance vector $\mathcal{R}(z)_{xy}$ as the Kronecker product

$$\mathcal{R}^T(z)_{xy} = \langle I_x^T(z) \otimes I_y^{*T}(z) \rangle$$
$$= [\mathcal{R}_0 \ \mathcal{R}_{01} \cdots \mathcal{R}_{0,N-1} \ \mathcal{R}_{10} \ \mathcal{R}_1 \cdots \mathcal{R}_{1,N-1} \ \cdots$$
$$\cdots \mathcal{R}_{N-1,0} \cdots \mathcal{R}_{N-1}]_{xy}, \quad (3.96)$$

where each element of this vector represents one of the functions of Equation (3.95) with abbreviated notation. Following the analysis of Section 3.4:

1. Replace the subscript x in Equation (3.94) by the subscript y; take the complex conjugate of the result.
2. Take the Kronecker product of Equation (3.94) with the result of step 1.
3. Take the expected value of the result.
4. Substitute the definition of Equations (3.95) and (3.96) for the resulting column vectors.
5. Substitute Equations (3.39) and (3.51).
6. Take the limit as $\Delta z \to 0$.

The calculations are a straightforward extension to those of Equations (3.59) and 3.60). We obtain

$$\mathcal{R}'(z)_{xy} = \{S_0(\mathbf{C} \otimes \mathbf{C} - \tfrac{1}{2}\mathcal{I} \otimes \mathbf{C}^2 - \tfrac{1}{2}\mathbf{C}^2 \otimes \mathcal{I}) \\ - \mathcal{I} \otimes \boldsymbol{\Gamma}_y^* - \boldsymbol{\Gamma}_x \otimes \mathcal{I}\} \cdot \mathcal{R}(z)_{xy}. \quad (3.97)$$

The present results for $\sigma = 0$ or $x = y$, in the two-mode or multi-mode cases respectively, yield the coupled power equations of Section 3.4. Treatment of general frequency-dependent C_{ij}, α_i, and β_i is straightforward by the present methods [3]. Both co- and cross-variances of the transfer functions are obtained by this analysis. We do not assume that different transfer functions are uncorrelated [2, 4]. Simplifications occur in the nondegenerate case (Section 4.6.2).

3.7. IMPULSE RESPONSE STATISTICS

The analysis of Section 3.6 for transfer-function statistics was carried out in the frequency domain. We interpret these results in the time domain for the signal-mode transfer function of the idealized two-mode guide of Section 3.2, with frequency-independent attenuation α_1 and α_2, coupling coefficient C_0, and group velocities v_1 and v_2.

44 GUIDES WITH WHITE RANDOM COUPLING

Consider a causal filter with real impulse response $g(t)$ and transfer function $\mathcal{G}(f)$:

$$g(t) = \int_{-\infty}^{\infty} \mathcal{G}(f) e^{j2\pi ft} df. \tag{3.98}$$

$$g(t) = g^*(t); \qquad g(t) = 0, \quad t < 0. \tag{3.99}$$

$$\mathcal{G}(f) = \int_{0}^{\infty} g(t) e^{-j2\pi ft} dt. \tag{3.100}$$

$$\mathcal{G}(f) = \mathcal{G}^*(-f). \tag{3.101}$$

Assume the transfer function is wide-sense stationary:

$$\mathcal{R}(\nu) = \langle \mathcal{G}(f+\nu) \mathcal{G}^*(f) \rangle; \qquad \mathcal{R}(\nu) = \mathcal{R}^*(-\nu). \tag{3.102}$$

$$\langle \mathcal{G}(f) \rangle = \langle \mathcal{G} \rangle. \tag{3.103}$$

The expected value and covariance of the transfer function are independent of the frequency f. The spectral density of the transfer function is $\mathcal{P}(-t)$, where $\mathcal{P}(t)$ is the Fourier transform of $\mathcal{R}(\nu)$:

$$\mathcal{P}(t) = \int_{-\infty}^{\infty} \mathcal{R}(\nu) e^{j2\pi t\nu} d\nu; \qquad \mathcal{P}(t) = \mathcal{P}^*(t). \tag{3.104}$$

Cascade the transfer function $\mathcal{G}(f)$ with an ideal band-pass filter of bandwidth B, to yield an overall transfer function $\mathcal{H}(f)$ as follows:

$$\mathcal{H}(f) = \begin{cases} \mathcal{G}(f), & |f - f_0| < B, \\ 0, & |f - f_0| > B, \end{cases} \quad B < f_0. \tag{3.105}$$

Let the impulse response of the overall filter be $h(t)$, with envelope

$$r_h(t) = |h(t) + j\hat{h}(t)|. \tag{3.106}$$

$\hat{h}(t)$ is the Hilbert transform of $h(t)$:

$$\hat{h}(t) = \frac{1}{\pi t} \star h(t) = \frac{1}{\pi} \int_{-\infty}^{\infty} \frac{h(\tau)}{t - \tau} d\tau, \tag{3.107}$$

3.7. IMPULSE RESPONSE STATISTICS

where $\star$ denotes the convolution operator. Then, it is shown in Section E.1 that

$$\langle \nu_h^2(t) \rangle \approx 8B\mathcal{P}(t), \qquad B \text{ large}, \qquad (3.108)$$

where B, the half-bandwidth of the ideal bandpass filter cascaded with $\mathcal{G}(f)$, is large compared to the correlation bandwidth of $\mathcal{G}(f)$ [1].

To summarize, $\mathcal{P}(-t)$ is the spectral density of the transfer function $\mathcal{G}(f)$. $\mathcal{P}(t)$ is proportional to the expected squared envelope of the impulse response of this transfer function, for large bandwidth at a high carrier frequency.

Let the filter input be the modulated white noise $\mathcal{I}(t)x(t)$, where $x(t)$ is a white noise of unit power:

$$\langle x(t+\tau)x(t) \rangle = \delta(\tau). \qquad (3.109)$$

$\mathcal{I}(t)$ is a real deterministic pulse; we call $\mathcal{I}^2(t)$ the input intensity. The filter output is

$$y(t) = [\mathcal{I}(t)x(t)] \star h(t) = \int_{-\infty}^{\infty} \mathcal{I}(\tau)x(\tau)h(t-\tau)d\tau. \qquad (3.110)$$

The output expected squared envelope is given in Section E.1 as

$$\langle \nu_y^2(t) \rangle = \mathcal{I}^2(t) \star \langle \nu_h^2(t) \rangle \approx 8B\mathcal{I}^2(t) \star \mathcal{P}(t), \qquad B \text{ large}. \qquad (3.111)$$

We call Equation (3.111) the output intensity, Equation (3.108) the intensity impulse response. A linear stationary random transfer function is linear also in intensity [4].

These results apply to the normalized impulse response of Section 3.2 by straightforward change of variables. The Fourier transforms of Equations (3.90) and (3.91) are [1]

$$P_0(\tau) = \int_{-\infty}^{\infty} \mathcal{R}_0(\sigma)e^{-j\tau\sigma z} d\left(\frac{\sigma z}{2\pi}\right) = e^{-S_0 z}\delta(\tau) + P_{0ac}(\tau), \qquad (3.112)$$

where

$$P_{0ac}(\tau) = \begin{cases} S_0 z \cdot e^{-S_0 z} e^{2\Delta\alpha z\tau} \sqrt{\frac{1-\tau}{\tau}} \mathbf{I}_1(2S_0 z\sqrt{\tau(1-\tau)}), & 0 < \tau < 1, \\ 0, & \text{otherwise.} \end{cases}$$

$$(3.113)$$

$$P_1(\tau) = \begin{cases} S_0 z \cdot e^{-S_0 z} e^{2\Delta\alpha z\tau} I_0(2S_0 z\sqrt{\tau(1-\tau)}), & 0 < \tau < 1. \\ 0, & \text{otherwise.} \end{cases} \quad (3.114)$$

$I_0(\ )$ and $I_1(\ )$ represent modified Bessel functions of orders 0 and 1, respectively.

The first term of Equation (3.112) is the spectral density of the d.c. component, and $P_{0ac}(-\tau)$ is the spectral density of the a.c. component of the transfer function $G_0(\Delta\beta)$, as discussed in Section E.1. The coefficient of $\delta(\tau)$ in Equation (3.112) is the square of $\langle G_0 \rangle$ in Equation (3.33). Moreover, $\mathcal{P}(t)$ of Equations (3.104) and (3.108) is related to $P_0(t)$ and $P_1(t)$ of Equations (3.112) and (3.114) by

$$\mathcal{P}(t) = \frac{1}{\Delta T} P_{0,1}\left(\frac{t}{\Delta T}\right), \quad (3.115)$$

where ΔT is given in Equation (3.20). We call $P_0(\tau)$ the normalized signal intensity impulse response. In Equation (3.112) $e^{-S_0 z} \cdot \delta(\tau)$ corresponds to the undistorted signal output. $P_{0ac}(\tau)$ corresponds to the undesired signal-mode output due to the random white coupling; we call $P_{0ac}(\tau)$ the normalized echo power. $P_1(\tau)$ is the normalized signal-spurious intensity impulse response.[3]

Finally, a direct analysis of the coupled line equations in the time domain, for the idealized two-mode case of Section 3.2, is given in Section E.2. Limiting forms for some of the present results have been obtained directly in the time domain [4].

3.8. DISCUSSION

We have obtained exact results for a number of transmission statistics of the coupled line equations with white coupling. This analysis contains no approximations of any kind; it can treat coherent and incoherent, single and multi-mode inputs. The methods are straightforward, and may readily be applied to similar problems; as one example, white random propagation constants are treated in Chapter 5.

---

[3] The usage of $P(\ )$ for normalized intensity impulse response differs from similar notation used to represent mode powers elsewhere in this text.

Closed-form analytical results are given above for two modes. A large number of modes requires the use of one of the symbolic computer algebra programs. Chapter 4 is devoted to applications of the present results, using MAPLE where required for the more complex examples. Cross-powers and sometimes higher mixed moments appear in this work; but Section 4.6 demonstrates substantial simplification in the absence of degenerate modes.

An open guide, such as an optical fiber or the dielectric slab waveguide of Appendix F, possesses radiation modes, in addition to the guided modes studied here. We ignore radiation modes throughout the present treatment; it has been suggested that their effect may be included in a phenomenological way by an appropriate contribution to the attenuation constant of each propagating mode.

The present results for white coupling can be extended to general coupling spectra under weak restrictions, by combining perturbation theory of Chapter 2 with the matrix techniques of the present chapter. This generalization is done in Chapter 6 for the coupled power equations. These results are validated by comparison with exact results for random square-wave coupling in Chapter 8. The extended results agree with the exact results of the present chapter for the limiting case of white coupling, and show the utility of the present results for almost white coupling spectra.

A number of related studies are contained in references [5–8].

REFERENCES

1. Harrison E. Rowe and D. T. Young, "Transmission Distortion in Multimode Random Waveguides," *IEEE Transactions on Microwave Theory and Techniques*, Vol. MTT-20, June 1972, pp. 349–365.
2. Dietrich Marcuse, *Theory of Dielectric Optical Waveguides*, 2nd ed., Academic Press, New York, 1991.
3. D. T. Young and Harrison E. Rowe, "Optimum Coupling for Random Guides with Frequency-Dependent Coupling," *IEEE Transactions on Microwave Theory and Techniques*, Vol. MTT-20, June 1972, pp. 365–372.
4. S. D. Personick, "Time Dispersion in Dielectric Waveguides," *Bell System Technical Journal*, Vol. 50, March 1971, pp. 843–859.
5. S. E. Miller, "Waveguide as a Communication Medium," *Bell System Technical Journal*, Vol. 33, November 1954, pp. 1209–1265.

6. J. A. Morrison and J. McKenna, "Coupled Line Equations with Random Coupling," *Bell System Technical Journal*, Vol. 51, January 1972, pp. 209–228.
7. S. D. Personick, "Two Derivations of the Time-Dependent Coupled-Power Equations," *Bell System Technical Journal*, Vol. 54, January 1975, pp. 47–52.
8. Richard Steinberg, "Pulse Propagation in Multimode Fibers with Frequency-Dependent Coupling," *IEEE Transactions on Microwave Theory and Techniques*, Vol. MTT-23, January 1975, pp. 121–122.

CHAPTER FOUR

Examples—White Coupling

4.1. INTRODUCTION

We give a number of applications of the results of Chapter 3. This study is not intended to be exhaustive, but rather to illustrate the range of possibilities offered by the present analysis. White coupling and constant propagation parameters are assumed, as described in Section 3.1. In addition, all modes are assumed lossless in the following examples, to reduce the number of parameters. We will consider coherent and incoherent, single- and multi-mode inputs for an N-mode guide, with $N = 2$ and $N = 4$.

For two-mode guide closed-form expressions are evaluated for the average mode powers and cross-powers, and for the second-order impulse response statistics. Power fluctuations are determined by MAPLE. The number of parameters describing the system is small enough to permit presentation of the results in general terms.

For more than two modes, there are too many parameters for a general treatment. We therefore consider a particular 4-mode guide described in Appendix F, and use MAPLE to obtain the average mode powers.

Finally, the nondegenerate case, in which all modes have distinct propagation constants, yields much simpler approximate results. These approximate results are readily modified to include degenerate modes, while retaining substantial simplification from the general exact results if most modes are nondegenerate.

The following inputs are used for these calculations.

4.1.1. Single-Mode Input

Consider first an input signal I_0 of unit power, with other mode inputs zero. All modes except the signal are zero at the input:

$$I_i(0) = 0, \quad 1 \le i \le N - 1. \tag{4.1}$$

For a coherent signal, i.e., a unit sinusoidal input,

$$I_0(0) = 1. \tag{4.2}$$

We consider as an incoherent signal a sinusoid with unit amplitude and random phase:[1]

$$I_0(0) = e^{j\theta_0}. \tag{4.3}$$

$$w(\theta_0) = \begin{cases} \dfrac{1}{2\pi}, & 0 < \theta_0 < 2\pi. \\ 0, & \text{otherwise.} \end{cases} \tag{4.4}$$

Then,

$$\langle I_0(0) \rangle = 0; \quad \langle |I_0(0)|^2 \rangle = 1. \tag{4.5}$$

$$\langle I_i(0)I_j(0) \rangle = \langle |I_i(0)|^2 |I_j(0)|^2 \rangle = \langle I_i^2(0)I_j^{*2}(0) \rangle = 0, \\ i \ne j, \ 0 \le i, j \le N - 1. \tag{4.6}$$

For both cases, coherent and incoherent, the average input powers and cross-powers are[2]

$$\mathcal{P}_0(0) = 1, \quad \mathcal{P}_i(0) = \langle P_i(0) \rangle = 0; \quad 1 \le i \le N - 1. \tag{4.7}$$

$$\mathcal{P}_{ij}(0) = \langle P_{ij}(0) \rangle = 0, \quad i \ne j, \ 0 \le i, j \le N - 1. \tag{4.8}$$

The higher moments are

$$\langle P_0^2(0) \rangle = 1, \quad \langle P_i(0)P_j(0) \rangle = \langle P_{ij}^2(0) \rangle = 0; \\ i \ne j, \ 0 \le i, j \le N - 1. \tag{4.9}$$

---

[1] Probability densities are denoted by $w(\ )$.
[2] The mode powers $P_i(z)$ are defined in Equations (2.3) and (2.46), the cross-powers $P_{ij}(z)$ and the average mode powers $\mathcal{P}_i(z)$ and cross-powers $\mathcal{P}_{ij}(z)$ in Equations (3.50).

4.1.2. Multi-Mode Coherent Input

A multi-mode coherent input, such as a plane wave, will excite all modes with unit correlation coefficients at the input. We assume unit sinusoidal inputs with zero phase for each of the N modes at the input:

$$I_i(0) = 1, \qquad 0 \leq i \leq N - 1. \tag{4.10}$$

The corresponding average powers and cross-powers at the input are

$$\mathcal{P}_i(0) = \langle P_i(0) \rangle = 1, \qquad 0 \leq i \leq N - 1. \tag{4.11}$$

$$\mathcal{P}_{ij}(0) = \langle P_{ij}(0) \rangle = 1, \qquad i \neq j,\ 0 \leq i, j \leq N - 1. \tag{4.12}$$

The higher moments are

$$\langle P_i^2(0) \rangle = \langle P_i(0) P_j(0) \rangle = \langle P_{ij}^2(0) \rangle = 1, \qquad 0 \leq i, j \leq N - 1. \tag{4.13}$$

4.1.3. Multi-Mode Incoherent Input

Finally, we consider a multi-mode incoherent input that excites unit power in each mode, with all modes statistically independent. We assume the input waves are unit sinusoids with independent, uniformly distributed random phase:

$$I_i(0) = e^{j\theta_i}, \qquad 0 \leq i \leq N - 1. \tag{4.14}$$

$$w(\theta_i) = \begin{cases} \dfrac{1}{2\pi}, & 0 < \theta_i < 2\pi. \\ 0, & \text{otherwise.} \end{cases} \tag{4.15}$$

$$w(\theta_1, \theta_2, \cdots \theta_{N-1}) = w(\theta_1) w(\theta_2) \cdots w(\theta_{N-1}). \tag{4.16}$$

Then,

$$\langle I_i(0) \rangle = 0, \quad \langle |I_i(0)|^2 \rangle = \langle |I_i(0)|^4 \rangle = 1; \qquad 0 \leq i \leq N - 1. \tag{4.17}$$

$$\langle |I_i(0)|^2 |I_j(0)|^2 \rangle = 1, \quad \langle I_i(0) I_j^*(0) \rangle = \langle I_i^2(0) I_j^{*2}(0) \rangle = 0;$$
$$i \neq j, \quad 0 \leq i, j \leq N - 1. \tag{4.18}$$

The input power and cross-power statistics are

$$\mathcal{P}_i(0) = \langle P_i(0) \rangle = 1, \quad \mathcal{P}_{ij}(0) = \langle P_{ij}(0) \rangle = 0;$$
$$i \neq j, \quad 0 \leq i, j \leq N - 1. \tag{4.19}$$

Higher moments become

$$\langle P_i^2(0) \rangle = 1, \quad \langle P_i(0) P_j(0) \rangle = 1, \quad \langle P_{ij}^2(0) \rangle = 0;$$
$$i \neq j, \quad 0 \leq i, j \leq N - 1. \tag{4.20}$$

4.2. COUPLED POWER EQUATIONS—TWO-MODE CASE

4.2.1. Single-Mode Input

Consider Equations (3.4)–(3.6) for two coupled modes; assume zero attenuation for both:

$$\alpha_0 = \alpha_1 = 0. \tag{4.21}$$

Assume input only to the signal mode; i.e., from Equation (4.1) with $N = 2$,

$$I_1(0) = 0. \tag{4.22}$$

The expected responses are given by Equations (3.32)–(3.33). For a coherent signal input, Equation (4.2):

$$\mathcal{I}_0(z) = \langle I_0(z) \rangle = e^{-j\beta_0 z} e^{-\frac{S_0}{2} z}, \quad \mathcal{G}_0(z) = \langle G_0(z) \rangle = e^{-\frac{S_0}{2} z}.$$
$$\mathcal{I}_1(z) = \langle I_1(z) \rangle = 0, \quad \mathcal{G}_1(z) = \langle G_1(z) \rangle = 0. \tag{4.23}$$

S_0 is given by Equations (3.1)–(3.2) as the spectral density of the (white) coupling coefficient. The spurious mode has no coherent component; the signal-mode coherent component decays exponentially to zero at large distance z, the rate of decay increasing with S_0.

For an incoherent signal input, Equations (4.3)–(4.4), neither signal nor spurious mode has a coherent component.

4.2. COUPLED POWER EQUATIONS—TWO-MODE CASE

The coupled power equations for the lossless case are, from Equations (3.53) and (4.21),

$$\begin{aligned} \mathcal{P}'_0(z) &= -S_0 \mathcal{P}_0(z) + S_0 \mathcal{P}_1(z), \\ \mathcal{P}'_1(z) &= S_0 \mathcal{P}_0(z) - S_0 \mathcal{P}_1(z). \end{aligned} \quad (4.24)$$

The initial conditions, from Equations (4.7), are the same for both coherent and incoherent signal-mode inputs; for $N = 2$,

$$\mathcal{P}_0(0) = 1, \quad \mathcal{P}_1(0) = 0. \quad (4.25)$$

The solutions to these equations are

$$\begin{aligned} \mathcal{P}_0(z) &= \frac{1}{2} + \frac{1}{2} e^{-2S_0 z}, \\ \mathcal{P}_1(z) &= \frac{1}{2} - \frac{1}{2} e^{-2S_0 z}. \end{aligned} \quad (4.26)$$

The results are plotted in Figure 4.1.

The cross-powers are given by Equation (3.54); in the lossless case they become

$$\begin{aligned} \mathcal{P}'_{01}(z) &= -(j\Delta\beta + S_0)\mathcal{P}_{01}(z) + S_0 \mathcal{P}_{10}(z), \\ \mathcal{P}'_{10}(z) &= S_0 \mathcal{P}_{01}(z) - (-j\Delta\beta + S_0)\mathcal{P}_{10}(z), \\ \Delta\beta &= \beta_0 - \beta_1, \end{aligned} \quad (4.27)$$

with initial conditions obtained from Equation (4.8) with $N = 2$ as

$$\mathcal{P}_{01}(0) = \mathcal{P}_{10}(0) = 0. \quad (4.28)$$

Therefore,

$$\mathcal{P}_{01}(z) = \mathcal{P}_{10}(z) = 0. \quad (4.29)$$

For signal only input the two modes are uncorrelated at the input, and remain so for all z; this is true for both coherent and incoherent signal mode inputs.

For a coherent signal input with zero spurious mode input, Equations (4.1) and (4.2), the signal mode contains a coherent component given by Equation (4.23), plus a random component. The spurious

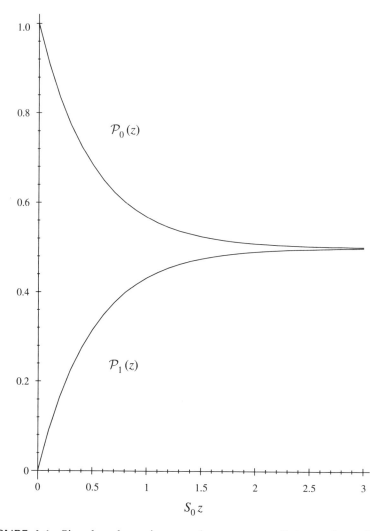

FIGURE 4.1. Signal and spurious mode powers vs. distance, for coherent or incoherent signal input.

mode contains only a random component. The signal power is therefore divided between coherent and incoherent components, the former given by $|\langle I_0(z)\rangle|^2$. We have

$$\mathcal{P}_0(z)_{\text{coherent}} = e^{-S_0 z}; \quad \mathcal{P}_0(z)_{\text{incoherent}} = \frac{1}{2} + \frac{1}{2}e^{-2S_0 z} - e^{-S_0 z}. \quad (4.30)$$

Figure 4.2 shows the total signal power and its two components.

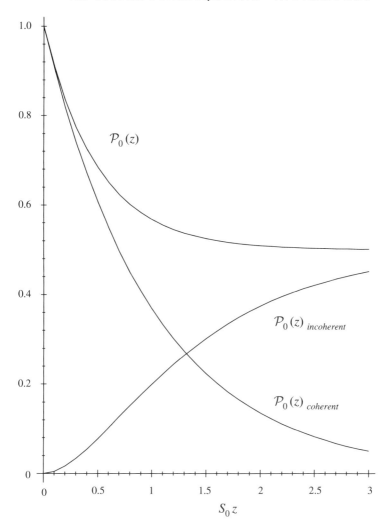

FIGURE 4.2. Coherent and incoherent signal mode powers vs. distance, for coherent signal input.

As z increases the coherent signal component decays exponentially to zero; for large z the total power divides equally between signal and spurious modes. Further discussion is given in Section 4.4.

4.2.2. Two-Mode Coherent Input

Consider next the input conditions of Equations (4.10)–(4.12) for $N = 2$. For the lossless case, Equation (4.21), the expected responses given by Equations (3.30)–(3.31) become

$$\mathcal{I}_0(z) = \langle I_0(z) \rangle = e^{-j\beta_0 z} e^{-\frac{S_0}{2} z}; \qquad \mathcal{G}_0(z) = \langle G_0(z) \rangle = e^{-\frac{S_0}{2} z}.$$
$$\mathcal{I}_1(z) = \langle I_1(z) \rangle = e^{-j\beta_1 z} e^{-\frac{S_0}{2} z}; \qquad \mathcal{G}_1(z) = \langle G_1(z) \rangle = e^{-\frac{S_0}{2} z}. \qquad (4.31)$$

Signal and spurious mode coherent components have equal magnitude, both decaying to zero at large distance z, with decay constant $S_0/2$.

Equation (4.24), the lossless coupled power equations, now have the trivial solutions

$$\mathcal{P}_0(z) = \mathcal{P}_1(z) = 1. \qquad (4.32)$$

These total powers are again composed of coherent and incoherent components, the former given by the squared magnitudes of the expected responses given in Equation (4.31). The incoherent components are

$$\mathcal{P}_0(z)_{\text{incoherent}} = \mathcal{P}_1(z)_{\text{incoherent}} = 1 - e^{-S_0 z}. \qquad (4.33)$$

The cross-powers are given by Equation (4.27), with initial conditions from Equation (4.12) with $N = 2$:

$$\mathcal{P}_{01}(0) = \mathcal{P}_{10}(0) = 1. \qquad (4.34)$$

The solutions are

$$\mathcal{P}_{01}(z) = \mathcal{P}_{10}^*(z) = \frac{e^{-S_0 z}}{2\sqrt{1 - \left(\frac{\Delta\beta}{S_0}\right)^2}}$$
$$\cdot \left\{ \left[\sqrt{1 - \left(\frac{\Delta\beta}{S_0}\right)^2} - 1 + j\left(\frac{\Delta\beta}{S_0}\right) \right] e^{-S_0 \sqrt{1 - \left(\frac{\Delta\beta}{S_0}\right)^2} z} \right. \qquad (4.35)$$
$$\left. + \left[\sqrt{1 - \left(\frac{\Delta\beta}{S_0}\right)^2} + 1 - j\left(\frac{\Delta\beta}{S_0}\right) \right] e^{+S_0 \sqrt{1 - \left(\frac{\Delta\beta}{S_0}\right)^2} z} \right\}.$$

The following special cases are of interest. For equal propagation constants,

$$\mathcal{P}_{01}(z) = \mathcal{P}_{10}(z) = 1, \qquad \Delta\beta = 0; \qquad (4.36)$$

this result is readily verified by elementary means. Next,

$$\mathcal{P}_{01}(z) = \mathcal{P}_{10}^*(z) = [1 + (1 - j) \cdot S_0 z] e^{-S_0 z}, \qquad \Delta\beta = S_0. \qquad (4.37)$$

4.2. COUPLED POWER EQUATIONS—TWO-MODE CASE

Finally, for large $\Delta\beta$ we have

$$\lim_{\Delta\beta\to\infty} |\mathcal{P}_{01}(z)| = \lim_{\Delta\beta\to\infty} |\mathcal{P}_{10}(z)| = e^{-S_0 z}. \quad (4.38)$$

Figure 4.3 shows the behavior of the cross-power magnitudes for several values of $(\Delta\beta/S_0)$. The two modes become uncorrelated for sufficiently large z. The four-mode guide described in Section 4.5

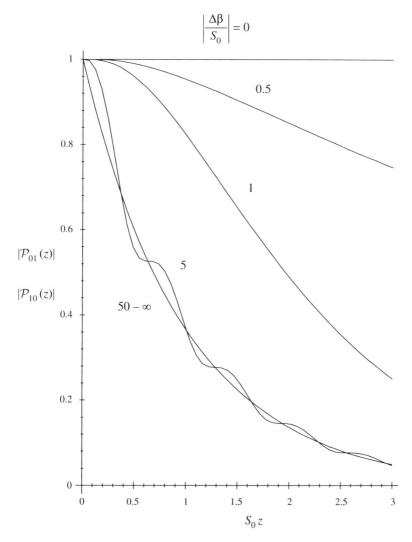

FIGURE 4.3. Cross-powers vs. distance, for coherent signal and spurious mode inputs.

has $\Delta\beta/S_0 \approx 10^{+25}$. This suggests that for non-degenerate modes only the $\Delta\beta/S_0 = \infty$ curve in Figure 4.3 will be of interest.

4.2.3. Two-Mode Incoherent Input

Consider finally the input conditions of Equations (4.14)–(4.19), with $N = 2$. For the lossless case, Equations (3.30)–(3.31), (4.24), and (4.27) yield the following trivial solutions:

$$\mathcal{I}_i(z) = \langle I_i(z) \rangle = 0, \quad \mathcal{G}_i(z) = \langle G_i(z) \rangle = 0; \quad i = 1, 2. \quad (4.39)$$

$$\mathcal{P}_0(z) = \mathcal{P}_1(z) = 1. \quad (4.40)$$

$$\mathcal{P}_{01}(z) = \mathcal{P}_{10}(z) = 0. \quad (4.41)$$

There is no coherent component in either signal or spurious mode. Both modes have constant average powers. Finally, the cross-powers are zero; consequently the two modes remain uncorrelated for all z.

4.3. POWER FLUCTUATIONS—TWO-MODE GUIDE

Next, we study the power fluctuations for two lossless coupled modes. The fourth-order statistics for this case are given by Equations (3.73)–(3.78). For convenience, introduce the following notation in Equations (3.73) and (3.74):

$$\mathcal{Q}^T(z) = [\mathcal{Q}_0(z) \quad \mathcal{Q}_1(z) \quad \mathcal{Q}_2(z) \quad \mathcal{Q}_3(z) \quad \mathcal{Q}_4(z)]. \quad (4.42)$$

$$\begin{aligned}
\mathcal{Q}_0(z) &= \langle P_0^2(z) \rangle. \\
\mathcal{Q}_1(z) &= \langle P_0(z) P_1(z) \rangle. \\
\mathcal{Q}_2(z) &= \langle P_{01}^2(z) \rangle. \\
\mathcal{Q}_3(z) &= \langle P_{10}^2(z) \rangle = \langle P_{01}^2(z) \rangle^*. \\
\mathcal{Q}_4(z) &= \langle P_1^2(z) \rangle.
\end{aligned} \quad (4.43)$$

The averages in Equation (4.43) are, respectively, the signal power second moment, the cross-moment of signal and spurious mode powers, the two cross-power second moments, and the spurious power second moment.

$\mathcal{Q}(z)$ is given by the solution of Equation (3.75),

$$\mathcal{Q}'(z) = \{S_0 \mathbf{N}_1 - \mathbf{N}_2\} \cdot \mathcal{Q}(z), \tag{4.44}$$

where $\mathbf{N}_1$ is given by Equation (3.76),

$$\mathbf{N}_1 = \begin{bmatrix} -2 & 4 & -1 & -1 & 0 \\ 1 & -4 & 1 & 1 & 1 \\ -1 & 4 & -2 & 0 & -1 \\ -1 & 4 & 0 & -2 & -1 \\ 0 & 4 & -1 & -1 & -2 \end{bmatrix}, \tag{4.45}$$

and $\mathbf{N}_2$ is given by Equation (3.77) and (3.78) for the lossless case as

$$\mathbf{N}_2 = \begin{bmatrix} 0 & 0 & 0 & 0 & 0 \\ 0 & 0 & 0 & 0 & 0 \\ 0 & 0 & -j2\Delta\beta & 0 & 0 \\ 0 & 0 & 0 & +j2\Delta\beta & 0 \\ 0 & 0 & 0 & 0 & 0 \end{bmatrix}. \tag{4.46}$$

The signal and spurious mode power variances are given by Equation (3.68):

$$\begin{aligned} \langle [\Delta P_0(z)]^2 \rangle &= \mathcal{Q}_0(z) - \langle P_0(z) \rangle^2 = \mathcal{Q}_0(z) - \mathcal{P}_0^2(z). \\ \langle [\Delta P_1(z)]^2 \rangle &= \mathcal{Q}_4(z) - \langle P_1(z) \rangle^2 = \mathcal{Q}_4(z) - \mathcal{P}_1^2(z). \end{aligned} \tag{4.47}$$

The corresponding standard deviations are

$$\begin{aligned} \Delta \mathcal{P}_0(z) &= \sqrt{\langle [\Delta P_0(z)]^2 \rangle} = \sqrt{\mathcal{Q}_0(z) - \mathcal{P}_0^2(z)}, \\ \Delta \mathcal{P}_1(z) &= \sqrt{\langle [\Delta P_1(z)]^2 \rangle} = \sqrt{\mathcal{Q}_4(z) - \mathcal{P}_1^2(z)}, \end{aligned} \tag{4.48}$$

where $\mathcal{P}_0(z)$ and $\mathcal{P}_1(z)$ are given in Section 4.2 for various inputs.

60 EXAMPLES—WHITE COUPLING

The initial conditions $\mathcal{Q}^T(0)$ for Equations (4.44) are found by substituting the relations of Sections 4.1.1–4.1.3 into Equations (4.42)–(4.43) with $z = 0$. These differential equations are solved by MAPLE, and the standard deviations for the signal and spurious mode power fluctuations are determined by Equation (4.48). $\Delta \mathcal{P}_0(z)$ and $\Delta \mathcal{P}_1(z)$ have the same units as the average powers, and indicate the range of statistical fluctuation in the mode powers $P_0(z)$ and $P_1(z)$ about their average values $\mathcal{P}_0(z)$ and $\mathcal{P}_1(z)$ due to random coupling.

$\mathcal{Q}_1(z)$, $\mathcal{Q}_2(z)$, and $\mathcal{Q}_3(z)$ are not of principal interest here.

4.3.1. Single-Mode Input

The initial conditions $\mathcal{Q}^T(0)$ are given by

$$\mathcal{Q}^T(0) = [1 \quad 0 \quad 0 \quad 0 \quad 0]. \tag{4.49}$$

This relation corresponds to zero spurious mode input and a constant amplitude input signal of unit power, either with constant phase (coherent) or random phase (incoherent). $\Delta \mathcal{P}_0(z)$ and $\Delta \mathcal{P}_1(z)$ are shown in Figure 4.4.

The mode power second moments in the degenerate two-mode case, $|\Delta \beta / S_0| = 0$ in Figure 4.4, are

$$\mathcal{Q}_0(z) = \langle P_0^2(z) \rangle = \frac{3}{8} + \frac{1}{8} e^{-8S_0 z} + \frac{1}{2} e^{-2S_0 z};$$
$$\mathcal{Q}_4(z) = \langle P_1^2(z) \rangle = \frac{3}{8} + \frac{1}{8} e^{-8S_0 z} - \frac{1}{2} e^{-2S_0 z}; \qquad \Delta \beta = 0. \tag{4.50}$$

The variances are

$$[\Delta \mathcal{P}_0(z)]^2 = [\Delta \mathcal{P}_1(z)]^2 = \frac{1}{8} + \frac{1}{8} e^{-8S_0 z} - \frac{1}{4} e^{-4S_0 z}, \qquad \Delta \beta = 0. \tag{4.51}$$

4.3.2. Two-Mode Coherent Input

The initial conditions for unit coherent inputs to signal and spurious modes are

$$\mathcal{Q}^T(0) = [1 \quad 1 \quad 1 \quad 1 \quad 1]. \tag{4.52}$$

$\Delta \mathcal{P}_0(z)$ and $\Delta \mathcal{P}_1(z)$ are shown in Figure 4.5. These two quantities are identically zero in the degenerate case, $\Delta \beta = 0$.

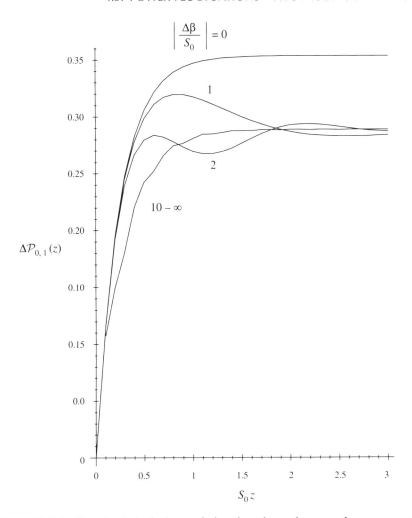

FIGURE 4.4. Standard deviations of signal and spurious mode powers vs. distance, with differential propagation constant as a parameter, for coherent or incoherent signal input.

4.3.3. Two-Mode Incoherent Input

Finally, the initial conditions for unit incoherent inputs are

$$\mathcal{Q}^T(0) = [1 \ \ 1 \ \ 0 \ \ 0 \ \ 1]. \tag{4.53}$$

$\Delta \mathcal{P}_0(z)$ and $\Delta \mathcal{P}_1(z)$ are plotted in Figure 4.6. In the degenerate case

$$\Delta \mathcal{P}_0(z) = \Delta \mathcal{P}_1(z) = .5\sqrt{1 - e^{-8S_0 z}}, \qquad \Delta \beta = 0. \tag{4.54}$$

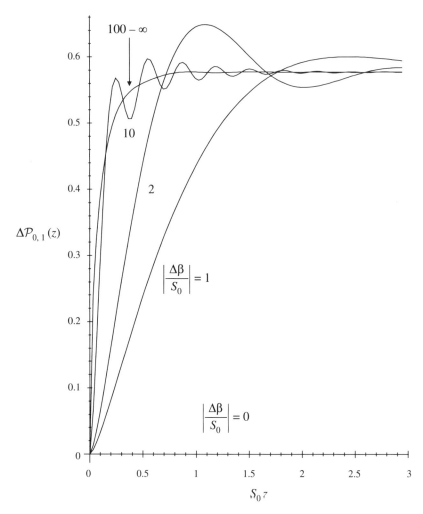

FIGURE 4.5. Standard deviations of signal and spurious mode powers vs. distance, with differential propagation constant as a parameter, for coherent signal and spurious mode inputs.

4.3.4. Discussion

The power fluctuations in the signal and spurious modes are equal in each of the three cases of Figures 4.4–4.6. This is required by conservation of power, Equation (2.5) in the lossless case. For large z, the mode powers have standard deviations comparable to their average values.

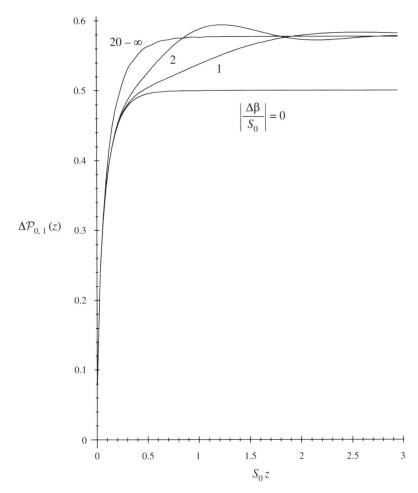

FIGURE 4.6. Standard deviations of signal and spurious mode powers vs. distance, with differential propagation constant as a parameter, for incoherent signal and spurious mode inputs.

The power fluctuations depend on $|\Delta\beta/S_0|$ for small values of this parameter. However, the example of Section 4.5 suggests that only the $|\Delta\beta/S_0| = \infty$ curves in these three figures are significant for nondegenerate modes.

The closed-form results given above for two degenerate modes, obtained by MAPLE, are significant in that they are readily obtained directly from the coupled line equations by elementary meth-

4.4. IMPULSE RESPONSE—TWO-MODE CASE

We next study some of the properties of the normalized signal-mode intensity impulse response of a lossless, two-mode guide with white coupling. From Equations (3.112) and (3.113),

$$P_0(\tau) = e^{-S_0 z}\delta(\tau) + P_{0ac}(\tau), \tag{4.55}$$

where

$$P_{0ac}(\tau) = \begin{cases} S_0 z \cdot e^{-S_0 z}\sqrt{\dfrac{1-\tau}{\tau}}I_1\left[2S_0 z\sqrt{\tau(1-\tau)}\right], & 0 \le \tau < 1. \\ 0, & \text{otherwise.} \end{cases} \tag{4.56}$$

$P_0(t)$ is proportional to the expected value of the squared envelope of the normalized impulse response of Equations (3.16)–(3.17). Recall that the real-time intensity impulse response $\mathcal{P}(t)$ for a dispersionless guide is obtained from Equations (4.55)–(4.56) by Equation (3.115), and that $\mathcal{P}(t)$ convolved with the input intensity $\mathcal{I}^2(t)$, is proportional to the corresponding output intensity (i.e., the real-time output expected squared envelope) by Equation (3.111). Equation (4.56) is plotted in Figure 4.7.

The normalized transfer function $G_0(\Delta\beta)$ is wide-sense stationary by Equations (3.90) and (3.93), and therefore satisfies Equations (E.1)–(E.15); it is the sum of a deterministic component $\langle G_0 \rangle = \overline{G_0} = e^{-\frac{S_0}{2}z}$ and a random component $G_{0ac}(\Delta\beta)$. Consequently, the first term of Equation (4.55), $e^{-S_0 z}\delta(\tau)$, represents the coherent part of the response, due to the deterministic part of the transfer function. In Equation (4.56), $P_{0ac}(\tau)$ represents the incoherent part of the response, due to the random part of the transfer function, previously denoted the normalized echo power.

For zero coupling, $S_0 z = 0$, $P_{0ac}(\tau) = 0$, and $P_0(\tau) = \delta(\tau)$. The input signal travels undistorted, entirely in the signal mode, at velocity $v_0 = 2\pi f/\beta_0$ of Equations (3.19)–(3.20); since $I_0(z) = e^{-j\beta_0 z}G_0(z)$,

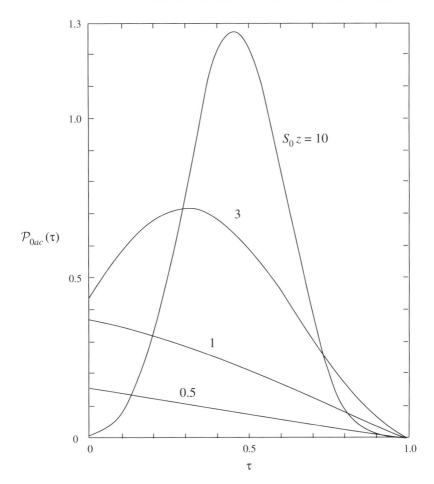

FIGURE 4.7. Normalized echo power, with coupling spectral density as a parameter. $\Delta\alpha = 0$. (From Harrison E. Rowe and D. T. Young, "Transmission Distortion in Multimode Random Waveguides," *IEEE Transactions on Microwave Theory and Techniques*, Vol. MTT-20, June 1972, pp. 349–365. © 1972 IEEE.)

the output arrives at time $L/v_0 = \beta_0 L/(2\pi f)$ corresponding to $\tau = 0$ in Figure 4.7.

For finite coupling, portions of the signal travel in both signal and spurious modes, at velocities v_0 and v_1, respectively, $v_0 > v_1$. A signal coupled into the spurious mode at $z = 0$, traveling to $z = L$ in the spurious mode, and coupled back into the signal mode at $z = L$, will arrive at the output at time $L/v_1 = \beta_1 L/(2\pi f)$, corresponding

66 EXAMPLES—WHITE COUPLING

to $\tau = 1$ in Figure 4.7. Generally, portions of the signal will travel in both modes, with arrival times corresponding to values $0 < \tau < 1$ in Figure 4.7.

We examine these results for small and for large coupling, to illustrate the physics of these results. Using the small-argument approximation for the modified Bessel function in Equation (4.56):

$$P_{0ac}(\tau) \approx (S_0 z)^2 e^{-S_0 z}(1 - \tau), \quad S_0 z \ll 1; \quad 0 \leq \tau < 1. \quad (4.57)$$

Using the asymptotic approximation for the Bessel function, and LaPlace's method:

$$P_{0ac}(\tau) \approx \sqrt{\frac{S_0 z}{2\pi}} \exp\left[-2S_0 z\left(\tau - \frac{1}{2}\right)^2\right], \quad S_0 z \gg 1; \quad 0 < \tau < 1.$$
(4.58)

Equation (4.57) may be obtained directly from perturbation theory, Equations (2.35) and (2.39), corresponding to the $n = 1$ term of Equations (B.1)–(B.2). By Equation (E.33) this approximation includes only all pairs of transitions, from signal to spurious mode and back. Since there are more pairs close together than far apart, the normalized echo power $P_{0ac}(\tau)$ in Equation (4.57) is triangular, as shown in Figure 4.7 for $S_0 z = 0.5$.

Within the regime of perturbation theory, perhaps $S_0 z < 0.5$, Equations (4.55) and (4.57) or Figure 4.7 show that the larger the coupling, the worse the performance; as S_0 increases the coherent component (i.e., the undistorted signal) decreases and the incoherent component (i.e., the echo distortion) increases. Conventional wisdom originally, and incorrectly, indicated that this behavior continued beyond the perturbation region. S. D. Personick first demonstrated that for very large coupling, increasing the coupling improved the performance [1]. This remarkable and unexpected result is demonstrated by Equation (4.58) and Figure 4.7. The total area under $P_{0ac}(\tau)$ for large coupling is from Equation (4.58)

$$\int_0^1 P_{0ac}(\tau) dt \approx \sqrt{\frac{S_0 z}{2\pi}} \int_{-\infty}^{\infty} e^{-2S_0 z \tau^2} dt = 0.5, \quad S_0 z \gg 1. \quad (4.59)$$

Therefore Equation (4.58) represents a narrow pulse of area 0.5 centered on $\tau = 0.5$, approximating $0.5\delta(\tau)$ for sufficiently large $S_0 z$.

As an example, for $S_0 z = 628$, $P_{0ac}(\tau)$ has a peak value $P_{0ac}(0.5) \approx 10$ and width ≈ 0.032 at the 36.8% points, where $P_{0ac}(0.5 \pm 0.016) \approx 3.68$; the coherent component has become insignificant.

As S_0 increases, more terms of the summation in Equation (B.1) become important. The partial impulse response corresponding to Equation (B.2) is given in Equation (E.33), which must be integrated over all possible sets of $2n$ transitions; for the great majority of these the signal will have spent about half of the time traveling in the signal and in the spurious modes. Consequently, the output will be a narrow pulse containing half the input power, with delay corresponding to the average velocity of signal and spurious modes. This behavior is described by the results of Equations (4.58) and (4.59).

4.5. COUPLED POWER EQUATIONS—FOUR-MODE CASE

We apply the results of Sections 3.3 and 3.4, for the average transfer functions and for the coupled power equations, to four lossless coupled modes. The large number of parameters necessary—6 coupling coefficients, 4 propagation constants, and the spectral density of the geometric parameter—do not allow a compact treatment of the general case, as was done for the two-mode case. Consequently, we consider as a specific example TE modes in a symmetric slab waveguide, with particular numerical parameters for dielectric constants, thickness, and frequency/wavelength given in Appendix F. The average transfer functions are given by Equation (3.41) as

$$\mathcal{I}'(z) = \left(-\Gamma - \frac{1}{2} S_0 \mathbf{C}^2\right) \cdot \mathcal{I}(z), \quad \mathcal{I}(z) = \langle I(z) \rangle. \tag{4.60}$$

The coupled power equations are from Equation (3.61)

$$\mathcal{P}'(z) = \left\{ S_0 \left(\mathbf{C} \otimes \mathbf{C} - \frac{1}{2} \mathcal{I} \otimes \mathbf{C}^2 - \frac{1}{2} \mathbf{C}^2 \otimes \mathcal{I} \right) - \mathcal{I} \otimes \Gamma^* - \Gamma \otimes \mathcal{I} \right\} \cdot \mathcal{P}(z). \tag{4.61}$$

The dielectric slab is illustrated in Figures F.1 and F.3, with index 1.01, thickness $1.23 \cdot 10^{-5}$ m, and wavelength 10^{-6} m, given in Equations (F.14) and (F.17). The propagation and coupling matrices are given by Equations (F.20) and (F.21) as follows:

$$\mathbf{\Gamma} = j \times 10^6 \begin{bmatrix} 6.3424 & 0 & 0 & 0 \\ 0 & 6.3316 & 0 & 0 \\ 0 & 0 & 6.3142 & 0 \\ 0 & 0 & 0 & 6.2921 \end{bmatrix} \text{m}^{-1}. \quad (4.62)$$

$$\mathbf{C} = 10^9 \begin{bmatrix} 0 & 2.3225 & 0 & 4.4966 \\ 2.3225 & 0 & 6.8737 & 0 \\ 0 & 6.8737 & 0 & 13.3083 \\ 4.4966 & 0 & 13.3083 & 0 \end{bmatrix} \text{m}^{-2}. \quad (4.63)$$

With these substitutions Equation (4.60) yields four coupled differential equations for the expected transfer functions, while Equation (4.61) yields 16 coupled differential equations for the average mode powers and cross-powers.

S_0 of Equations (4.60) and (4.61), defined in Equations (3.1)–(3.2), is the spectral density of $c(z)$ in Equation (3.9). In the present case, $c(z)$ is the straightness deviation of the dielectric slab illustrated in Figure F.3. We need to select a numerical value for S_0 consistent with a physical model of this guide.

The present white model for $c(z)$ has infinite straightness deviation for any finite value of S_0. The mean-square straightness deviation for a general coupling spectral density $S(\nu)$ is

$$\langle c^2(z) \rangle = \int_{-\infty}^{\infty} S(\nu) d\nu. \quad (4.64)$$

We show in Chapter 6 that the present results, for white coupling spectra, hold under reasonable restrictions for ideal low-pass $S(\nu)$ that cover the range of spatial frequencies $0 < \nu < \nu_c$, where

$$\nu_c \gg \frac{|\beta_i - \beta_j|_{\max}}{2\pi}, \quad \text{all } i, j. \quad (4.65)$$

In the present case, Equation (4.62) requires that $S(\nu)$ be flat up to the spatial frequency

$$\nu_c \gg \frac{|\beta_0 - \beta_3|}{2\pi} = \frac{0.0503 \times 10^6}{2\pi} \approx 8 \cdot 10^3 \text{m}^{-1}. \quad (4.66)$$

4.5. COUPLED POWER EQUATIONS—FOUR-MODE CASE

We somewhat arbitrarily choose the following parameters for $S(\nu)$:

$$S(\nu) = \begin{cases} S_0 = 10^{-21} \text{ m}^3, & |\nu| < 2 \times 10^4 \text{ m}^{-1}. \\ 0, & |\nu| > 2 \times 10^4 \text{ m}^{-1}. \end{cases} \quad (4.67)$$

From Equation (4.64), this corresponds to an rms straightness deviation of

$$\sqrt{\langle c^2(z) \rangle} \approx 6.325 \times 10^{-9} \text{ m}, \quad (4.68)$$

small compared to the guide width in Equation (F.17) of $1.23 \cdot 10^{-5}$ m.

We use MAPLE to solve Equations (4.60) and (4.61) for the numerical parameters of Equations (4.62), (4.63), and (4.67), for three representative inputs.

4.5.1. Single-Mode Input

The initial conditions for unit signal input are given in Section 4.1.1. For coherent input, we have for Equation (4.60)

$$\mathcal{I}^T(0) = \langle I^T(0) \rangle = [1 \ 0 \ 0 \ 0]. \quad (4.69)$$

The magnitude of the average signal transfer function is shown in Figure 4.8.

Modes 1 and 3 have zero coherent component

$$\langle I_1(z) \rangle = \langle I_3(z) \rangle = 0. \quad (4.70)$$

While the expected values of modes 0 and 2 are coupled, in contrast to the two-mode case of Section 4.2.1, the coupling is very weak:

$$|\langle I_2(z) \rangle| < 3 \times 10^{-5}, \quad (4.71)$$

too small to appear in Figure 4.8. The average signal transfer function appears to decay exponentially, similar to the two-mode case.

For incoherent signal input, all four modes have zero coherent component.

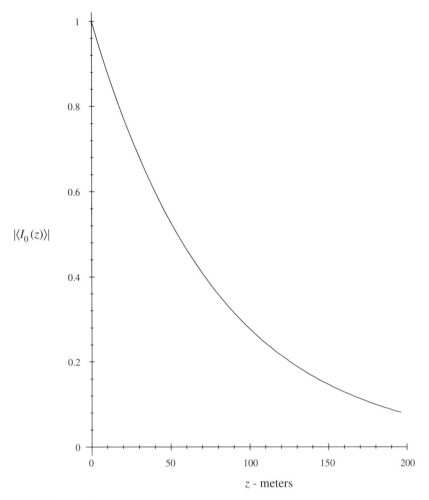

FIGURE 4.8. Average signal transfer function magnitude, for coherent single-mode input.

The initial conditions for Equation (4.61) for unit coherent or incoherent signal input are

$$\mathcal{P}^{T}(0) = \begin{bmatrix} 1 & \underbrace{0 \ 0 \ 0 \ 0 \ 0 \ 0 \ 0 \ 0 \ 0 \ 0 \ 0 \ 0 \ 0 \ 0 \ 0}_{15 \text{ zeros}} \end{bmatrix}. \tag{4.72}$$

The average mode powers for the four modes are shown in Figure 4.9.

4.5. COUPLED POWER EQUATIONS—FOUR-MODE CASE

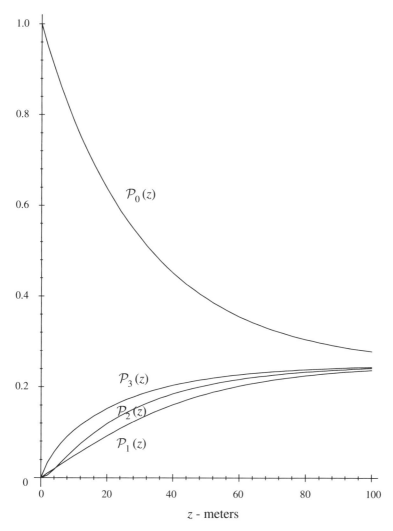

FIGURE 4.9. Mode powers vs. distance, for coherent or incoherent signal input.

For large z, the total power divides equally between all four modes. If the input signal is coherent, $\mathcal{P}_0(z)$ contains a coherent component $\langle I_0(z) \rangle$, whose magnitude is shown in Figure 4.8, with power given by $|\langle I_0(z) \rangle|^2$. Modes 1 and 3 have zero coherent component; the coherent component of mode 2 is too small to appear on the scale of Figure 4.9.

4.5.2. Multi-Mode Coherent Input

The initial conditions for four-mode coherent input are given in Section 4.1.2. For Equation (4.60),

$$\mathcal{I}^T(0) = \langle I^T(0) \rangle = [1 \quad 1 \quad 1 \quad 1]. \tag{4.73}$$

The magnitudes of the expected values of the four modes are shown in Figure 4.10.

Each mode appears to decay exponentially, with its own decay constant. Further study shows that the coupling between all four expected mode amplitudes is weak, as in the preceding section.

The initial conditions for the coupled power equations, Equation (4.61), for four-mode coherent input are

$$\mathcal{P}^T(0) = [\underbrace{1 \quad 1 \quad 1 \quad 1 \quad 1 \quad 1 \quad 1 \quad 1 \quad 1 \quad 1 \quad 1 \quad 1 \quad 1 \quad 1 \quad 1 \quad 1}_{16 \text{ ones}}]. \tag{4.74}$$

The average powers, $\mathcal{P}_0(z)$, $\mathcal{P}_1(z)$, $\mathcal{P}_2(z)$, and $\mathcal{P}_3(z)$, depart only infinitesimally from their initial values of 1, by less than $\pm 4 \times 10^{-6}$ over the range $0 < z < 30$, and imperceptibly for greater z. Each contains a coherent component, given by the square magnitude of the corresponding average mode amplitude, shown on Figure 4.10.

4.5.3. Multi-Mode Incoherent Input

For four-mode incoherent input, we have the following initial conditions from Section 4.1.3:

$$\mathcal{I}^T(0) = \langle I^T(0) \rangle = [0 \quad 0 \quad 0 \quad 0]. \tag{4.75}$$

$$\mathcal{P}^T(0) = [1 \quad 0 \quad 0 \quad 0 \quad 0 \quad 1 \quad 0 \quad 0 \quad 0 \quad 0 \quad 1 \quad 0 \quad 0 \quad 0 \quad 0 \quad 1]. \tag{4.76}$$

Equations (4.75) and (4.60) yield

$$\langle I_0(z) \rangle = \langle I_1(z) \rangle = \langle I_2(z) \rangle = \langle I_3(z) \rangle = 0; \tag{4.77}$$

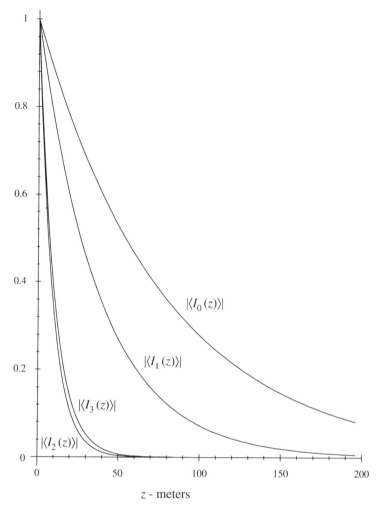

FIGURE 4.10. Average mode amplitudes, for four-mode coherent input.

all four modes have zero coherent component. Equations (4.76) and (4.61) yield

$$\mathcal{P}_0(z) = \mathcal{P}_1(z) = \mathcal{P}_2(z) = \mathcal{P}_3(z) = 1. \tag{4.78}$$

For large z the average mode powers are equal, with zero coherent components. Since the initial conditions satisfy this asymptotic behavior, the coupled power equations yield no further change in the average powers.

4.6. NONDEGENERATE CASE—APPROXIMATE RESULTS

The two-mode case of Section 4.2 exhibits simplifications not present in the four-mode results of Section 4.5:

1. The expected mode amplitudes are exponential, and uncoupled.
2. The average mode powers are uncoupled from the cross-powers.

Nevertheless, in this four-mode case the expected mode amplitudes appear at least approximately exponential, and are coupled only very weakly. Moreover, Marcuse has presented multi-mode coupled power equations in which cross-powers do not appear [2]; it will appear in Section 4.6.2 below that his simpler relations, obtained from 4 coupled differential equations, give results indistinguishable from those of Figure 4.9, obtained from 16 coupled differential equations. These facts suggest further simplifications of the present general results in special cases.

The approximations that we desire arise from the disparate numerical values of the coefficients in the differential equations for the average mode amplitudes and average powers, Equations (4.60) and (4.61). In the particular example of Section 4.5, with numerical values given by Equations (4.62), (4.63), and (4.67), the first term on the right-hand side of Equation (4.60) is of order $O(10^{+6})$, the second $O(10^{-3})$; the first term on the right-hand side of Equation (4.61) is $O(10^{-3})$, the second and third terms $O(10^{+6})$. The principal contributions to the derivatives on the left-hand sides of these two differential equations thus arise from the terms involving Γ in this particular case. This suggests, as a first step, removing the deterministic portion of the solutions involving the propagation constants, by the change of variables given in Equation (3.11):

$$I(z) = e^{-\Gamma z} \cdot G(z). \tag{4.79}$$

Then find the asymptotic solutions as all the components of Γ, i.e., the mode propagation constants $\Gamma_0, \Gamma_1, \Gamma_2, \cdots$ of Equation (3.10), approach infinity. In the nondegenerate case no two Γ_i's will be equal, and all of the differential propagation constants $\Delta\Gamma_{ij}$ will also approach infinity. Special account must be taken for degenerate modes, for which $\Delta\Gamma = 0$.

4.6. NONDEGENERATE CASE—APPROXIMATE RESULTS

4.6.1. Average Transfer Functions

Substituting Equation (4.79) into Equation (4.60), we find

$$\langle G(z) \rangle' = -\frac{1}{2} S_0 \left(e^{+\Gamma z} \cdot \mathbf{C}^2 \cdot e^{-\Gamma z} \right) \cdot \langle G(z) \rangle. \qquad (4.80)$$

The coupling matrix $\mathbf{C}$ is given by Equation (2.45) or (3.35); the elements of $\mathbf{C}^2$ are as follows:[3]

$$(\mathbf{C}^2)_{ij} = \sum_{k=0}^{N-1} C_{i-1,k} C_{k,j-1} = \sum_{k=0}^{N-1} C_{i-1,k} C_{j-1,k}, \quad 1 \leq i, j \leq N. \qquad (4.81)$$

For diagonal elements, this becomes

$$(\mathbf{C}^2)_{ii} = \sum_{k=0}^{N-1} C_{i-1,k} C_{k,i-1} = \sum_{k=0}^{N-1} C_{i-1,k}^2, \quad 1 \leq i \leq N. \qquad (4.82)$$

Note that the $k = i - 1$ and $k = j - 1$ terms in Equation (4.81) are zero, while the $k = i - 1$ terms in Equation (4.82) are zero. Equation (3.10) yields

$$e^{\Gamma z} = e^{(\boldsymbol{\alpha}+j\boldsymbol{\beta})z} = \begin{bmatrix} e^{(\alpha_0+j\beta_0)z} & 0 & 0 & \cdots \\ 0 & e^{(\alpha_1+j\beta_1)z} & 0 & \cdots \\ 0 & 0 & e^{(\alpha_2+j\beta_2)z} & \cdots \\ \vdots & \vdots & \vdots & \ddots \end{bmatrix}. \qquad (4.83)$$

Now the diagonal terms of the matrix on the right-hand side of Equation (4.80) are identical to those of Equation (4.82):

$$\left(e^{+\Gamma z} \cdot \mathbf{C}^2 \cdot e^{-\Gamma z} \right)_{ii} = (\mathbf{C}^2)_{ii} = \sum_{k=0}^{N-1} C_{i-1,k}^2, \quad 1 \leq i \leq N. \qquad (4.84)$$

---

[3]The matrix elements are numbered from 1, 1 to N, N, while the nonzero coupling coefficients in Equation 2.45, with unequal subscripts, are numbered from 0, 1 to $N - 2, N - 1$.

However, for the off-diagonal terms the exponential factors do not cancel out; using Equations (4.81) and (4.83),

$$\left(e^{+\Gamma z} \cdot \mathbf{C}^2 \cdot e^{-\Gamma z}\right)_{ij} = e^{\Delta \Gamma_{i-1,j-1} z} (\mathbf{C}^2)_{ij}$$

$$= e^{\Delta \Gamma_{i-1,j-1} z} \sum_{k=0}^{N-1} C_{i-1,k} C_{k,j-1}, \quad 1 \le i, j \le N, \quad (4.85)$$

where

$$\Delta \Gamma_{ij} = \Gamma_i - \Gamma_j = \Delta \alpha_{ij} + j \Delta \beta_{ij} = (\alpha_i - \alpha_j) + j(\beta_i - \beta_j),$$
$$0 \le i, j \le N - 1. \quad (4.86)$$

Consider now the solution of Equation (4.80) by the method of successive approximations, similar to that of Appendix A. The right-hand-side integrals in each step of the successive approximations will approach zero for each of the off-diagonal terms in Equation (4.85), for $i \ne j$, as $\Delta \beta_{ij} \to \infty$, by the method of stationary phase [3]. Therefore in the asymptotic limit the differential equations of Equation (4.80) become uncoupled, and we have

$$\langle G_i(z) \rangle_\infty = \langle G_i(0) \rangle e^{-\frac{1}{2} S_0 z \sum_{k=0}^{N-1} C_{ik}^2}, \quad 0 \le i \le N - 1, \quad (4.87)$$

where the subscript ∞ indicates the asymptotic limit. Finally, substituting Equations (4.79) and (4.83),

$$\langle I_i(z) \rangle_\infty = \langle I_i(0) \rangle e^{-\Gamma_i z} e^{-\frac{1}{2} S_0 z \sum_{k=0}^{N-1} C_{ik}^2}, \quad 0 \le i \le N - 1. \quad (4.88)$$

The average transfer functions become uncoupled and exponential for large $\Delta \beta_{ij}$.

For the four-mode example of Section 4.5, with single-mode input given by Equation (4.69), Equation (4.88) becomes

$$|\langle I_0(z) \rangle|_\infty = e^{-0.01281 z}. \quad (4.89)$$

In this case, the exponential factor $e^{j \Delta \beta_{02} z} = e^{j(\beta_0 - \beta_2) z}$ has approximately 10^6 periods along the horizontal axis of Figure 4.8. Therefore the asymptotic approximation of Equation (4.89) is a good approximation to the exact solution shown in Figure 4.8, obtained by solving

4.6. NONDEGENERATE CASE—APPROXIMATE RESULTS

four simultaneous differential equations. The coefficient in the exponent for the ith mode in Equations (4.87) and (4.88) is the power per unit length coupled from the ith mode to all of the other modes; the $k = i$ term in the summation is zero.

4.6.2. Coupled Power Equations

We perform the substitution of Equation (4.79) in the coupled power equations, Equation (4.61). From Equations (2.43), (3.50), and (3.57),

$$P(z) = I(z) \otimes I^*(z), \tag{4.90}$$

$$\mathcal{P}(z) = \langle P(z) \rangle = \langle I(z) \otimes I^*(z) \rangle. \tag{4.91}$$

Substituting Equation (4.79) into Equation (4.90),

$$P(z) = \left[e^{-\Gamma z} \cdot G(z)\right] \otimes \left[e^{-\Gamma^* z} \cdot G^*(z)\right]$$
$$= \left[e^{-\Gamma z} \otimes e^{-\Gamma^* z}\right] \cdot \left[G(z) \otimes G^*(z)\right]. \tag{4.92}$$

We require the following additional notation:

$$\mathcal{R}_i(z) = \langle R_i(z) \rangle, \quad R_i(z) = |G_i(z)|^2. \tag{4.93}$$

$$\mathcal{R}_{ij}(z) = \mathcal{R}^*_{ji}(z) = \langle R_{ij}(z) \rangle, \quad R_{ij}(z) = G_i(z) G^*_j(z). \tag{4.94}$$

$$\mathcal{R}(z) = \langle R(z) \rangle, \quad R(z) = G(z) \otimes G^*(z). \tag{4.95}$$

$$\mathcal{R}^T(z) = \begin{bmatrix} \mathcal{R}_0 \ \mathcal{R}_{01} \cdots \mathcal{R}_{0,N-1} & \mathcal{R}_{10} \ \mathcal{R}_1 \cdots \mathcal{R}_{1,N-1} & \cdots \\ \cdots & \mathcal{R}_{N-1,0} \ \mathcal{R}_{N-1,1} \cdots \mathcal{R}_{N-1} \end{bmatrix}, \tag{4.96}$$

where the z dependence of the $\mathcal{R}_i$ and $\mathcal{R}_{ij}$ has been suppressed. Then, Equations (4.92) and (4.95) yield

$$P(z) = \left(e^{-\Gamma z} \otimes e^{-\Gamma^* z}\right) \cdot R(z). \tag{4.97}$$

Differentiating Equation (4.97),

$$P'(z) = \left(e^{-\Gamma z} \otimes e^{-\Gamma^* z}\right)' \cdot R(z) + \left(e^{-\Gamma z} \otimes e^{-\Gamma^* z}\right) \cdot R'(z). \tag{4.98}$$

Evaluating the first factor on the right-hand side,

$$\begin{aligned}(e^{-\Gamma z} \otimes e^{-\Gamma^* z})' \\ = (-\Gamma \cdot e^{-\Gamma z}) \otimes e^{-\Gamma^* z} + e^{-\Gamma z} \otimes (-\Gamma^* \cdot e^{-\Gamma^* z}) \\ = -(\Gamma \cdot e^{-\Gamma z}) \otimes (\mathcal{I} \cdot e^{-\Gamma^* z}) + (\mathcal{I} \cdot e^{-\Gamma z}) \otimes (-\Gamma^* \cdot e^{-\Gamma^* z}) \quad (4.99) \\ = -(\Gamma \otimes \mathcal{I}) \cdot (e^{-\Gamma z} \otimes e^{-\Gamma^* z}) - (\mathcal{I} \otimes \Gamma^*) \cdot (e^{-\Gamma z} \otimes e^{-\Gamma^* z}) \\ = -(\Gamma \otimes \mathcal{I} + \mathcal{I} \otimes \Gamma^*) \cdot (e^{-\Gamma z} \otimes e^{-\Gamma^* z}),\end{aligned}$$

where $\mathcal{I}$ is the unit matrix. Substituting Equation (4.99) into Equation (4.98), and taking the expected value using Equations (4.91) and (4.95),

$$\begin{aligned}\mathcal{P}'(z) = -(\Gamma \otimes \mathcal{I} + \mathcal{I} \otimes \Gamma^*) \cdot (e^{-\Gamma z} \otimes e^{-\Gamma^* z}) \cdot \mathcal{R}(z) \\ + (e^{-\Gamma z} \otimes e^{-\Gamma^* z}) \cdot \mathcal{R}'(z). \quad (4.100)\end{aligned}$$

Similarly, the expected value of Equation (4.97) yields

$$\mathcal{P}(z) = (e^{-\Gamma z} \otimes e^{-\Gamma^* z}) \cdot \mathcal{R}(z). \quad (4.101)$$

Finally, substituting Equations (4.100) and (4.101) into Equation (4.61), we obtain the desired result:

$$\begin{aligned}\mathcal{R}'(z) = S_0 (e^{\Gamma z} \otimes e^{\Gamma^* z}) \cdot \left(\mathbf{C} \otimes \mathbf{C} - \frac{1}{2}\mathcal{I} \otimes \mathbf{C}^2 - \frac{1}{2}\mathbf{C}^2 \otimes \mathcal{I}\right) \\ \cdot (e^{-\Gamma z} \otimes e^{-\Gamma^* z}) \cdot \mathcal{R}(z). \quad (4.102)\end{aligned}$$

Using the transformation of Equation (C.8), this may be written

$$\mathcal{R}'(z) = S_0 \left(\mathbf{M}_1 - \frac{1}{2}\mathbf{M}_2 - \frac{1}{2}\mathbf{M}_3\right) \cdot \mathcal{R}(z), \quad (4.103)$$

where

$$\mathbf{M}_1 = \mathbf{V} \otimes \mathbf{V}^*, \quad \mathbf{M}_2 = \mathcal{I} \otimes \mathbf{W}^*, \quad \mathbf{M}_3 = \mathbf{W} \otimes \mathcal{I}, \quad (4.104)$$

and

$$\mathbf{V} = e^{\Gamma z} \cdot \mathbf{C} \cdot e^{-\Gamma z}, \quad \mathbf{W} = e^{\Gamma z} \cdot \mathbf{C}^2 \cdot e^{-\Gamma z}. \quad (4.105)$$

We require the asymptotic limit of these results as $\Delta\beta_{ij} \to \infty$, $i \neq j$. Toward this end we divide $\mathcal{R}(z)$ into subvectors of dimension

4.6. NONDEGENERATE CASE—APPROXIMATE RESULTS

N as follows:

$$\mathcal{R}^T(z) = [\mathcal{R}_1^T(z) \quad \mathcal{R}_2^T(z) \quad \cdots \quad \mathcal{R}_N^T(z)]. \tag{4.106}$$

$$\begin{aligned}\mathcal{R}_i^T(z) = [\mathcal{R}_{i-1,0}(z) \quad \mathcal{R}_{i-1,1}(z) \quad \cdots \quad \mathcal{R}_{i,i-2}(z) \\ \mathcal{R}_{i-1}(z) \quad \mathcal{R}_{i-1,i}(z) \quad \cdots \quad \mathcal{R}_{i-1,N-1}(z)]. \end{aligned} \tag{4.107}$$

We divide each of the three $\mathbf{M}$ matrices into $N \times N$ submatrices:

$$\mathbf{M}_1 = \begin{bmatrix} \mathbf{M}_{1|11} & \mathbf{M}_{1|12} & \mathbf{M}_{1|13} & \cdots & \mathbf{M}_{1|1N} \\ \mathbf{M}_{1|21} & \mathbf{M}_{1|22} & \mathbf{M}_{1|23} & \cdots & \mathbf{M}_{1|2N} \\ \mathbf{M}_{1|31} & \mathbf{M}_{1|32} & \mathbf{M}_{1|33} & \cdots & \mathbf{M}_{1|3N} \\ \vdots & \vdots & \vdots & \ddots & \vdots \\ \mathbf{M}_{1|N1} & \mathbf{M}_{1|N2} & \mathbf{M}_{1|N3} & \cdots & \mathbf{M}_{1|NN} \end{bmatrix}, \tag{4.108}$$

with a similar partition for $\mathbf{M}_2$ and $\mathbf{M}_3$. Then the ith subvector of $\mathcal{R}(z)$ is found from Equations (4.103)–(4.107) as the solution of

$$\mathcal{R}_i'(z) = S_0 \sum_{k=1}^{N} \left(\mathbf{M}_{1|ik} - \frac{1}{2}\mathbf{M}_{2|ik} - \frac{1}{2}\mathbf{M}_{3|ik} \right) \cdot \mathcal{R}_k(z), \quad 1 \le i \le N. \tag{4.109}$$

Appendix H gives the asymptotic form for the $\mathbf{M}$ matrices and their submatrices.

First, substitute Equations (H.5)–(H.7) into the ith row of Equation (4.109):

$$\mathcal{R}'_{i-1}(z)_\infty = S_0 \sum_{k=1}^{N} C^2_{i-1,k-1}[e^{2\Delta\alpha_{i-1,k-1}z}\mathcal{R}_{k-1}(z)_\infty - \mathcal{R}_{i-1}(z)_\infty], \quad 1 \le i \le N. \tag{4.110}$$

Using Equations (4.79), (4.93), and (3.50), and relabeling the summation index in Equation (4.110), we have the final desired result:

$$\mathcal{P}_i'(z)_\infty = -\left(2\alpha_i + S_0 \sum_{k=0}^{N-1} C^2_{i,k}\right)\mathcal{P}_i(z)_\infty + S_0 \sum_{k=0}^{N-1} C^2_{i,k}\mathcal{P}_k(z)_\infty, \quad 0 \le i \le N-1. \tag{4.111}$$

The subscript ∞ again denotes the asymptotic limit as $\Delta\beta_{i,j} \to \infty$. Note from Equation (3.35) that the $k = i$ terms in the summations of Equation (4.111) are zero. As noted in Section 4.6.1, the quantity $S_0 \sum_{k=0}^{N-1} C_{i,k}^2$ is the power per unit length coupled from the ith mode to all of the other modes; $S_0 C_{i,k}^2 \mathcal{P}_k(z)_\infty$ is the power per unit length coupled from the kth mode to the ith mode.

The result of Equation (4.111) shows that for large $\Delta\beta_{i,j}$ the average power become uncoupled from the cross-powers. Marcuse's multi-mode coupled power equations [2], which contain only average powers but not cross-powers, are thus seen to represent the asymptotic limit for large $\Delta\beta_{ij}$. For the four-mode case of Section 4.5 Equations (4.111) become

$$\begin{bmatrix} \mathcal{P}_0'(z)_\infty \\ \mathcal{P}_1'(z)_\infty \\ \mathcal{P}_2'(z)_\infty \\ \mathcal{P}_3'(z)_\infty \end{bmatrix} = \begin{bmatrix} -0.0256 & 0.0054 & 0 & 0.0202 \\ 0.0054 & -0.0526 & 0.0472 & 0 \\ 0 & 0.0472 & -0.2243 & 0.1771 \\ 0.0202 & 0 & 0.1771 & -0.1973 \end{bmatrix} \cdot \begin{bmatrix} \mathcal{P}_0(z)_\infty \\ \mathcal{P}_1(z)_\infty \\ \mathcal{P}_2(z)_\infty \\ \mathcal{P}_3(z)_\infty \end{bmatrix}. \quad (4.112)$$

For single-mode input, with initial conditions

$$\mathcal{P}_0(0)_\infty = 1, \quad \mathcal{P}_1(0)_\infty = \mathcal{P}_2(0)_\infty = \mathcal{P}_3(0)_\infty = 0, \quad (4.113)$$

the solutions to the four simultaneous differential equations of Equation (4.112) are indistinguishable from those of Figure 4.9, obtained from the sixteen simultaneous differential equations of Equations (4.61)–(4.63) and (4.72).

Finally, consider rows other than the ith of Equation (4.109). For $i \neq j$ Equation (H.5) shows that $\mathbf{M}_{1|ik}$ makes zero contribution to this summation. Equations (H.6) and (H.7) yield

$$\mathcal{R}_{i-1,j-1}'(z)_\infty = -\frac{S_0}{2} \sum_{k=1}^{N} \left(C_{i-1,k-1}^2 + C_{j-1,k-1}^2 \right) \mathcal{R}_{i-1,j-1}(z)_\infty. \quad (4.114)$$

From Equations (3.50), (4.79), and (4.94),

$$|\mathcal{P}_{ij}(z)_\infty| = \exp\left[(\alpha_i + \alpha_j)z\right]\exp\left[-\frac{S_0}{2}\sum_{k=0}^{N-1}(C_{ik}^2 + C_{jk}^2)z\right]|\mathcal{P}_{ij}(0)_\infty|. \tag{4.115}$$

For the lossless two-mode case, Equation (3.42) yields the following analytic result for the $|\Delta\beta/S_0| = \infty$ curve of Figure (4.3):

$$|\mathcal{P}_{01}(z)_\infty| = |\mathcal{P}_{10}(z)_\infty| = e^{-S_0 z}. \tag{4.116}$$

This confirms the prior result of Equation (4.38), obtained directly as the limit of Equation (4.35).

4.6.3. Power Fluctuations

As an alternative to a general treatment of power fluctuations, we use MAPLE directly for the two-mode case. By substituting Equation (4.79) into Equations (3.64), (3.70), and (3.72), we have

$$\mathcal{R}_4'(z) = S_0 \left(e^{\Gamma z} \otimes e^{\Gamma^* z} \otimes e^{\Gamma z} \otimes e^{\Gamma^* z}\right)$$

$$\cdot \Big\{ \mathcal{I} \otimes \mathcal{I} \otimes \mathbf{C} \otimes \mathbf{C} - \mathcal{I} \otimes \mathbf{C} \otimes \mathcal{I} \otimes \mathbf{C} + \mathcal{I} \otimes \mathbf{C} \otimes \mathbf{C} \otimes \mathcal{I}$$

$$+ \mathbf{C} \otimes \mathcal{I} \otimes \mathcal{I} \otimes \mathbf{C} - \mathbf{C} \otimes \mathcal{I} \otimes \mathbf{C} \otimes \mathcal{I} + \mathbf{C} \otimes \mathbf{C} \otimes \mathcal{I} \otimes \mathcal{I}$$

$$- \frac{1}{2}(\mathcal{I} \otimes \mathcal{I} \otimes \mathcal{I} \otimes \mathbf{C}^2 + \mathcal{I} \otimes \mathcal{I} \otimes \mathbf{C}^2 \otimes \mathcal{I}$$

$$+ \mathcal{I} \otimes \mathbf{C}^2 \otimes \mathcal{I} \otimes \mathcal{I} + \mathbf{C}^2 \otimes \mathcal{I} \otimes \mathcal{I} \otimes \mathcal{I}) \Big\}$$

$$\cdot \left(e^{-\Gamma z} \otimes e^{-\Gamma^* z} \otimes e^{-\Gamma z} \otimes e^{-\Gamma^* z}\right) \cdot \mathcal{R}_4(z), \tag{4.117}$$

where

$$\mathcal{R}_4(z) = \left(e^{\Gamma z} \otimes e^{\Gamma^* z} \otimes e^{\Gamma z} \otimes e^{\Gamma^* z}\right) \cdot \mathcal{P}(z). \tag{4.118}$$

Taking the asymptotic limit of Equation (4.117) as $\Delta\beta_{ij} \to \infty$, $i \neq j$,

$$\mathcal{R}_4'(z)_\infty = S_0 \mathbf{M}_\infty \cdot \mathcal{R}_4(z)_\infty, \tag{4.119}$$

where $\mathbf{M}_\infty$ is found by setting to zero every complex term of the matrix on the right-hand side of Equation (4.117).

For the two-mode case, we have from Equation (3.42)

$$\mathbf{C} = \begin{bmatrix} 0 & 1 \\ 1 & 0 \end{bmatrix}, \quad \mathbf{C}^2 = \mathcal{I}. \quad (4.120)$$

Substituting Equation (4.120) into Equation (4.117) and taking the asymptotic limit, MAPLE yields for the lossless case

$$\mathbf{M}_\infty = \begin{bmatrix} -2 & 0 & 0 & 1 & 0 & 0 & 1 & 0 & 0 & 1 & 0 & 0 & 1 & 0 & 0 & 0 \\ 0 & -2 & 0 & 0 & -1 & 0 & 0 & 1 & 0 & 0 & 0 & 0 & 0 & 1 & 0 & 0 \\ 0 & 0 & -2 & 0 & 0 & 0 & 0 & 0 & -1 & 0 & 0 & 1 & 0 & 0 & 1 & 0 \\ 1 & 0 & 0 & -2 & 0 & 0 & -1 & 0 & 0 & -1 & 0 & 0 & 0 & 0 & 0 & 1 \\ 0 & -1 & 0 & 0 & -2 & 0 & 0 & 1 & 0 & 0 & 0 & 0 & 0 & 1 & 0 & 0 \\ 0 & 0 & 0 & 0 & 0 & -2 & 0 & 0 & 0 & 0 & 0 & 0 & 0 & 0 & 0 & 0 \\ 1 & 0 & 0 & -1 & 0 & 0 & -2 & 0 & 0 & 0 & 0 & -1 & 0 & 0 & 0 & 1 \\ 0 & 1 & 0 & 0 & 1 & 0 & 0 & -2 & 0 & 0 & 0 & 0 & -1 & 0 & 0 & 0 \\ 0 & 0 & -1 & 0 & 0 & 0 & 0 & 0 & -2 & 0 & 0 & 1 & 0 & 0 & 1 & 0 \\ 1 & 0 & 0 & -1 & 0 & 0 & 0 & 0 & 0 & -2 & 0 & 0 & -1 & 0 & 0 & 1 \\ 0 & 0 & 0 & 0 & 0 & 0 & 0 & 0 & 0 & 0 & -2 & 0 & 0 & 0 & 0 & 0 \\ 0 & 0 & 1 & 0 & 0 & 0 & 0 & 0 & 1 & 0 & 0 & -2 & 0 & 0 & -1 & 0 \\ 1 & 0 & 0 & 0 & 0 & 0 & -1 & 0 & 0 & -1 & 0 & 0 & -2 & 0 & 0 & 1 \\ 0 & 1 & 0 & 0 & 1 & 0 & 0 & -1 & 0 & 0 & 0 & 0 & 0 & -2 & 0 & 0 \\ 0 & 0 & 1 & 0 & 0 & 0 & 0 & 0 & 1 & 0 & 0 & -1 & 0 & 0 & -2 & 0 \\ 0 & 0 & 0 & 1 & 0 & 0 & 1 & 0 & 0 & 1 & 0 & 0 & 1 & 0 & 0 & -2 \end{bmatrix}$$

(4.121)

The 16 differential equations of Equations (4.119) and (4.121) separate into two independent sets of 8 equations each, as in Section 3.5. The set containing rows 1, 4, 7, 10, 13, and 16 yields the mode powers and their moments, of present interest. Rows 4, 7, 10, and 13 are equivalent. By eliminating this redundancy, and using Equations (4.118), (3.64), (3.66), and (3.70), we obtain 3 differential equations relating the fourth moments of the average mode powers for the

4.6. NONDEGENERATE CASE—APPROXIMATE RESULTS

lossless two-mode case:

$$\begin{bmatrix} \mathcal{Q}_0'(z)_\infty \\ \mathcal{Q}_1'(z)_\infty \\ \mathcal{Q}_4'(z)_\infty \end{bmatrix} = \begin{bmatrix} -2 & 4 & 0 \\ 1 & -4 & 1 \\ 0 & 4 & -2 \end{bmatrix} \cdot \begin{bmatrix} \mathcal{Q}_0(z)_\infty \\ \mathcal{Q}_1(z)_\infty \\ \mathcal{Q}_4(z)_\infty \end{bmatrix}, \quad (4.122)$$

where

$$\mathcal{Q}_0(z)_\infty = \langle P_0^2(z) \rangle_\infty.$$
$$\mathcal{Q}_1(z)_\infty = \langle P_0(z) P_1(z) \rangle_\infty. \quad (4.123)$$
$$\mathcal{Q}_4(z)_\infty = \langle P_1^2(z) \rangle_\infty.$$

Equation (4.122) represents the asymptotic limit of Equation (4.43), providing a valid approximation for large $\Delta\beta_{12}$.

Equations (4.122)–(4.123) yield analytic solutions for $|\Delta\beta/S_0| = \infty$. For single-mode input, Section 4.3.1, the initial conditions are

$$\mathcal{Q}_0(0)_\infty = 1, \qquad \mathcal{Q}_1(0)_\infty = 0, \qquad \mathcal{Q}_4(0)_\infty = 0, \quad (4.124)$$

and the solutions are

$$\Delta \mathcal{P}_0(z) = \Delta \mathcal{P}_1(z) = \sqrt{\frac{1}{12} - \frac{1}{4} e^{-4z} + \frac{1}{6} e^{-6z}}. \quad (4.125)$$

For two-mode coherent or incoherent input, Section 4.3.2 or 4.3.3, the initial conditions are

$$\mathcal{Q}_0(0)_\infty = 1, \qquad \mathcal{Q}_1(0)_\infty = 1, \qquad \mathcal{Q}_4(0)_\infty = 1, \quad (4.126)$$

with solutions

$$\Delta \mathcal{P}_0(z) = \Delta \mathcal{P}_1(z) = \sqrt{\frac{1}{3} - \frac{1}{3} e^{-6z}}. \quad (4.127)$$

Equations (4.125) and (4.127) yield the $|\Delta\beta/S_0| = \infty$ curves for Figures 4.4 and for Figures 4.5 and 4.6, respectively.

4.6.4. Discussion

The asymptotic approximation provides major simplification in the solutions for transfer function statistics of multi-mode transmission systems with white coupling. In the two-mode case, the differential equations for power fluctuations are reduced in number from 16 to 3, and provide a simple closed-form solution.

The physical example of Appendix F, a dielectric slab waveguide supporting four modes, suggests that this approximation will apply to many practical guides. Departures from these approximate results can occur only with strict degeneracy between one or more pairs of modes, i.e. with mode pairs having identical propagation constants. Under these special conditions additional non-zero terms in the various matrices are readily included, with corresponding modifications to the corresponding differential equations.

4.7. DISCUSSION

The examples of the present chapter illustrate the range of transmission statistics that can be obtained for multi-mode transmission systems. We assume that all modes travel in the same (forward) direction. The coupling spectrum is assumed white, i.e., the coupling (due to imperfections) is assumed to have a short correlation distance; this assumption permits exact results, without approximation of any kind. The further condition of distinct propagation constants (the nondegenerate case of Section 4.6) yields simplification, and excellent agreement with these exact results.

The present methods are indirect, in that the guide is divided into discrete sections as an intermediate step, permitting the use of Kronecker products, yielding a straightforward analysis. Kronecker product methods are used throughout the following chapters in treatments of other problems. We extend the present work to include arbitrary coupling spectra in Chapter 6.

Successively higher order statistics give increased information about the transfer functions of multi-mode guide. Average transfer functions and average mode powers provide no information about signal distortion, which first appears in the second-order (two-frequency) transfer function statistics or the equivalent time-domain result, i.e., the corresponding intensity impulse response.

The case of an N-mode guide with identical propagation constants for all modes is treated in Appendix G. This obviously nonphysical case is of interest in that it permits exact solution without using Kronecker products. These results provide confirmation of the present methods.

REFERENCES

1. S. D. Personick, "Time Dispersion in Dielectric Waveguides," *Bell System Technical Journal*, Vol. 50, March 1971, pp. 843–859.
2. Dietrich Marcuse, *Theory of Dielectric Optical Waveguides*, 2nd ed., Academic Press, New York, 1991.
3. Athanasios Papoulis, *The Fourier Integral and its Applications*, McGraw-Hill, New York, 1962.

CHAPTER FIVE

Directional Coupler with White Propagation Parameters

5.1. INTRODUCTION

We apply the coupled line equations for two forward-traveling modes, Equations (2.1) and (2.2), to the study of directional couplers with random imperfections. We repeat these equations for convenience:

$$I_0'(z) = -\Gamma_0(z)I_0(z) + jc(z)I_1(z),$$
$$I_1'(z) = jc(z)I_0(z) - \Gamma_1(z)I_1(z). \qquad (5.1)$$

$$\Gamma_0(z) = \alpha_0(z) + j\beta_0(z), \qquad \Gamma_1(z) = \alpha_1(z) + j\beta_1(z). \qquad (5.2)$$

An ideal directional coupler has constant coupling and constant propagation parameters. We assume in addition that the ideal coupler is lossless and symmetric; then Equations (5.1) and (5.2) become [1–5]

$$I_0'(z) = -j\beta I_0(z) + jcI_1(z),$$
$$I_1'(z) = jcI_0(z) - j\beta I_1(z). \qquad (5.3)$$

Backward-traveling modes are neglected. A typical coupler is shown schematically in Figure 5.1, with inputs

$$I_0(0) = 1, \qquad I_1(0) = 0. \qquad (5.4)$$

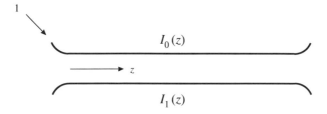

FIGURE 5.1. Directional coupler.

This sketch could represent identical parallel optical fibers, microstrips, etc.; the coupling c increases as the two guides are brought closer together. The solutions to Equations (5.3) with the inputs of Equations (5.4) are [6]

$$I_0(z) = \cos cz e^{-j\beta z}, \qquad I_1(z) = j \sin cz e^{-j\beta z}. \qquad (5.5)$$

Complete power transfer from mode 0 to mode 1 occurs for $z = \pi/(2c)$; this can occur only for a symmetric coupler, with identical propagation constants, and constant coupling, as in Equation (5.3).

A real coupler will have random variations in geometric and material parameters that will cause the coupling $c(z)$ and the propagation parameters $\Gamma_1(z)$ and $\Gamma_2(z)$ of Equations (5.1) and (5.2) to depart from their design values. For a symmetric coupler, $\Delta\beta(z) = \beta_0(z) - \beta_1(z)$ will vary randomly around zero, and $c(z)$ will vary randomly about c of Equation (5.3).

It is plausible that the variations in $\Delta\beta(z)$ will have the most significant effect on the transfer functions of the coupler. The statistical model of the following section assumes constant coupling and independent white propagation parameters for the two coupled guides [7]. The results of the present chapter are thus complementary to those of Chapters 3 and 4 for two-mode guides with white coupling and constant $\Delta\beta$.

5.2. STATISTICAL MODEL

Assume that the propagation parameters in Equations (5.1) and (5.2) are lossless, statistically independent, and stationary:

$$\begin{aligned} \beta_0(z) &= \beta + b_0(z). \\ \beta_1(z) &= \beta + b_1(z). \end{aligned} \qquad (5.6)$$

β is the common mean value of the two propagation constants, the design value of Equation (5.3):

$$\beta = \langle \beta_0(z) \rangle = \langle \beta_1(z) \rangle. \tag{5.7}$$

The a.c. components of the propagation parameters, $b_0(z)$ and $b_1(z)$, have zero mean, the same covariance, and zero cross-variance:

$$\langle b_0(z) \rangle = \langle b_1(z) \rangle = 0.$$

$$R_b(\zeta) = \langle b_0(z+\zeta)b_0(z) \rangle = \langle b_1(z+\zeta)b_1(z) \rangle. \tag{5.8}$$

$$\langle b_0(z+\zeta)b_1(z) \rangle = 0.$$

They have spectral density

$$B(\nu) = \int_{-\infty}^{\infty} R_b(\zeta) e^{-j2\pi\nu\zeta} d\zeta, \tag{5.9}$$

and zero cross-spectrum.

We assume the coupling coefficient of Equation (5.1) is constant, equal to its design value of Equation (5.3):

$$c(z) = c. \tag{5.10}$$

We find the average transfer functions and average powers for white propagation parameters:

$$R_b(\zeta) = B_0 \delta(\zeta), \qquad B(\nu) = B_0. \tag{5.11}$$

While the white assumption is nonphysical, in that it corresponds to an infinite variance for the propagation parameters, Chapter 6 shows that the results for white coupling are a good approximation for almost white coupling. We expect the present results for white $B(\nu)$ of Equation (5.11) will provide a good approximation for low-pass $B(\nu)$, under appropriate restrictions.

Let the bandwidth of $B(\nu)$ be denoted by ν_b, with the corresponding correlation length λ_b:

$$\lambda_b = \frac{1}{\nu_b}. \tag{5.12}$$

We divide the coupler into sections Δz. Then, the following assumptions must be satisfied for physical systems:

1. Independent sections.

$$\Delta z \gg \lambda_b = \frac{1}{\nu_b}. \tag{5.13}$$

2. Small phase shift per section, Equation (2.29).

$$\phi_k = \int_{(k-1)\Delta z}^{k\Delta z} \Delta\beta(z)dz. \tag{5.14}$$

Then following the footnote of Equation (3.28), from Equation (5.8)

$$\langle \phi_k^2 \rangle = \int_{(k-1)\Delta z}^{k\Delta z} [b_0^2(z) + b_1^2(z)]dz \approx 2B_0 \Delta z \ll 1. \tag{5.15}$$

3. Small coupling.

$$c\Delta z \ll 1. \tag{5.16}$$

4. Small fractional variation in $\beta_0(z)$, $\beta_1(z)$.

$$2B_0 \nu_b \ll \beta^2. \tag{5.17}$$

5. No coupling to backward modes.

$$\nu_b \ll \beta. \tag{5.18}$$

Equations (5.13) and (5.16) yield

$$c \ll \frac{1}{dz} \ll \nu_b. \tag{5.19}$$

Equations (5.13) and (5.15) yield

$$2B_0 \ll \frac{1}{dz} \ll \nu_b. \tag{5.20}$$

Equations (5.18) and (5.20) yield Equation (5.17).

Therefore the physical restrictions on the parameters of the following analysis based on the white spectrum assumed in Equation (5.11) for the a.c. components of the propagation parameters are as follows:

$$c \ll \frac{1}{dz} \ll \nu_b \ll \beta.$$
$$2B_0 \ll \frac{1}{dz} \ll \nu_b \ll \beta.$$
(5.21)

These conditions correspond to a symmetric coupler with imperfections that vary slowly compared to the optical wavelength, vary rapidly compared to the coupler length, and are sufficiently small that little relative phase shift occurs between the two modes in the correlation length of the imperfections.

5.3. AVERAGE TRANSFER FUNCTIONS

An elementary section of a lossless symmetric coupler is described by substituting Equations (5.6) and (5.10) into Equations (2.26)–(2.28):

$$\begin{bmatrix} \tilde{G}_0(k\Delta z) \\ \tilde{G}_1(k\Delta z) \end{bmatrix} = \begin{bmatrix} e^{-j\int_{(k-1)\Delta z}^{k\Delta z} b_0(z)dz} \cos c\Delta z & e^{-j\int_{(k-1)\Delta z}^{k\Delta z} b_1(z)dz} j \sin c\Delta z \\ e^{-j\int_{(k-1)\Delta z}^{k\Delta z} b_0(z)dz} j \sin c\Delta z & e^{-j\int_{(k-1)\Delta z}^{k\Delta z} b_1(z)dz} \cos c\Delta z \end{bmatrix}$$
$$\cdot \begin{bmatrix} \tilde{G}_0[(k-1)\Delta z] \\ \tilde{G}_1[(k-1)\Delta z] \end{bmatrix}.$$
(5.22)

We have defined for convenience the auxiliary quantities $\tilde{G}_0(z)$ and $\tilde{G}_1(z)$ as follows:

$$\begin{bmatrix} I_0(z) \\ I_1(z) \end{bmatrix} = e^{-j\beta z} \begin{bmatrix} \tilde{G}_0(z) \\ \tilde{G}_1(z) \end{bmatrix},$$
(5.23)

where the tildes distinguish these quantities from those of Equation (2.31). Define the expected values of the complex mode amplitudes as

$$\mathcal{I}_i(z) = \langle I_i(z) \rangle, \quad \tilde{\mathcal{G}}_i(z) = \langle \tilde{G}_i(z) \rangle; \quad i = 0, 1.$$
(5.24)

The different sections of Equation (5.22) are independent for white propagation parameters, Equation (5.11). Take the expected value of Equation (5.22), expand the matrix elements in Taylor series for small Δz, and use Equation (5.15):

$$\begin{bmatrix} \tilde{\mathcal{G}}_0(k\Delta z) \\ \tilde{\mathcal{G}}_1(k\Delta z) \end{bmatrix} = \begin{bmatrix} 1 - \dfrac{B_0}{2}\Delta z & jc\Delta z \\ jc\Delta z & 1 - \dfrac{B_0}{2}\Delta z \end{bmatrix} \cdot \begin{bmatrix} \tilde{\mathcal{G}}_0[(k-1)\Delta z] \\ \tilde{\mathcal{G}}_1[(k-1)\Delta z] \end{bmatrix}. \tag{5.25}$$

From Equations (D.8) and (D.14), for small Δz the difference equations of Equation (5.25) are approximated by the differential equations

$$\begin{bmatrix} \tilde{\mathcal{G}}_0'(z) \\ \tilde{\mathcal{G}}_1'(z) \end{bmatrix} = \begin{bmatrix} -\dfrac{B_0}{2} & jc \\ jc & -\dfrac{B_0}{2} \end{bmatrix} \cdot \begin{bmatrix} \tilde{\mathcal{G}}_0(z) \\ \tilde{\mathcal{G}}_1(z) \end{bmatrix}. \tag{5.26}$$

Assume the initial conditions of Equation (5.4) and Figure 5.1:

$$\mathcal{I}_0(0) = \tilde{\mathcal{G}}_0(0) = 1; \qquad \mathcal{I}_1(0) = \tilde{\mathcal{G}}_1(0) = 0. \tag{5.27}$$

Then the solutions to Equation (5.26) are

$$\begin{bmatrix} \tilde{\mathcal{G}}_0(z) \\ \tilde{\mathcal{G}}_1(z) \end{bmatrix} = e^{-(B_0/2)z} \begin{bmatrix} \cos cz \\ j\sin cz \end{bmatrix} \tag{5.28}$$

or, using Equations (5.23) and (5.24)

$$\begin{bmatrix} \mathcal{I}_0(z) \\ \mathcal{I}_1(z) \end{bmatrix} = e^{-[j\beta+(B_0/2)]z} \begin{bmatrix} \cos cz \\ j\sin cz \end{bmatrix}. \tag{5.29}$$

These results are an exponentially damped version of the results for an ideal coupler, given in Equation (5.5), the damping increasing as the variations in the propagation parameters increase. For $B_0 = 0$, Equation (5.29) is identical to Equation (5.5), with complete power transfer at $z = \pi/(2c)$. For an imperfect coupler, with

$B_0 > 0$, $\langle I_0[\pi/(2c)] \rangle = 0$, but this of course does *not* imply complete power transfer, since the angles $\angle I_0[\pi/(2c)]$ and $\angle I_1[\pi/(2c)]$ are random variables.

For zero coupling, $c = 0$, Equation (5.29) becomes

$$\begin{bmatrix} \mathcal{I}_0(z) \\ \mathcal{I}_1(z) \end{bmatrix} = e^{-[j\beta + (B_0/2)]z} \begin{bmatrix} 1 \\ 0 \end{bmatrix}. \tag{5.30}$$

For this case the solution to Equations (5.1), (5.2), (5.4), and (5.6) is

$$\begin{bmatrix} I_0(z) \\ I_1(z) \end{bmatrix} = e^{-j\beta z} \begin{bmatrix} e^{-j\int_0^z b_0(x)dx} \\ 0 \end{bmatrix}. \tag{5.31}$$

The exponent in Equation (5.31) will be Gaussian if $b_0(z)$ is Gaussian; this exponent will be approximately Gaussian in many other cases by the assumption of Equation (5.13). Then the expected value of Equation (5.31) follows directly from the characteristic function of a Gaussian random variable, yielding Equation (5.30). Thus, the expected complex mode amplitude of mode 0, $\langle I_0(z) \rangle$, decays exponentially because the total phase shift performs a random walk, with variance proportional to distance z; the power in mode 0 remains constant, $\langle |I_0(z)|^2 \rangle = 1$. The complex amplitude of mode 1 is identically zero, $I_1(z) = 0$, with expected value and power both zero.

5.4. COUPLED POWER EQUATIONS

The average powers and cross-powers are defined in Equation (3.50). Using this relation and Equation (5.23),

$$\mathcal{P}(z) = \begin{bmatrix} \mathcal{P}_0(z) \\ \mathcal{P}_{01}(z) \\ \mathcal{P}_{10}(z) \\ \mathcal{P}_1(z) \end{bmatrix} = \mathcal{I}(z) \otimes \mathcal{I}^*(z) = \tilde{\mathcal{G}}(z) \otimes \tilde{\mathcal{G}}^*(z), \tag{5.32}$$

where from Equation (5.24)

$$\mathcal{I}(z) = \begin{bmatrix} \mathcal{I}_0(z) \\ \mathcal{I}_1(z) \end{bmatrix} = \langle I(z) \rangle, \quad \tilde{\mathcal{G}}(z) = \begin{bmatrix} \tilde{\mathcal{G}}_0(z) \\ \tilde{\mathcal{G}}_1(z) \end{bmatrix} = \langle \tilde{G}(z) \rangle, \tag{5.33}$$

with

$$I(z) = \begin{bmatrix} I_0(z) \\ I_1(z) \end{bmatrix}, \quad \tilde{G}(z) = \begin{bmatrix} \tilde{G}_0(z) \\ \tilde{G}_1(z) \end{bmatrix}. \quad (5.34)$$

Substitute Equation (5.22), use Equations (D.15) and (D.17), expand the resulting matrix elements for small Δz, use Equation (5.15), and let $\Delta z \to 0$ to yield the coupled power equations:

$$\begin{bmatrix} \mathcal{P}'_0(z) \\ \mathcal{P}'_{01}(z) \\ \mathcal{P}'_{10}(z) \\ \mathcal{P}'_1(z) \end{bmatrix} = \begin{bmatrix} 0 & -jc & +jc & 0 \\ -jc & -B_0 & 0 & +jc \\ +jc & 0 & -B_0 & -jc \\ 0 & +jc & -jc & 0 \end{bmatrix} \cdot \begin{bmatrix} \mathcal{P}_0(z) \\ \mathcal{P}_{01}(z) \\ \mathcal{P}_{10}(z) \\ \mathcal{P}_1(z) \end{bmatrix}. \quad (5.35)$$

For two modes with white coupling, the powers and cross-powers in Equations (3.52)–(3.54) were decoupled. Here, for random propagation parameters, they are inextricably entwined.

The initial conditions corresponding to Equations (5.4) and (5.27) become

$$\begin{bmatrix} \mathcal{P}_0(0) \\ \mathcal{P}_{01}(0) \\ \mathcal{P}_{10}(0) \\ \mathcal{P}_1(0) \end{bmatrix} = \begin{bmatrix} 1 \\ 0 \\ 0 \\ 0 \end{bmatrix}. \quad (5.36)$$

Equations (5.35) and (5.36) yield the following results for the average powers and cross-powers:

$$\mathcal{P}_0(z) = 1 - \mathcal{P}_1(z)$$

$$= \frac{1}{2} + \frac{1}{2} e^{-(B_0/2)z} \left\{ \cos\left[2c\sqrt{1 - \left(\frac{B_0}{4c}\right)^2} \cdot z\right] \right.$$

$$\left. + \frac{B_0}{4c} \frac{\sin\left[2c\sqrt{1 - \left(\frac{B_0}{4c}\right)^2} \cdot z\right]}{\sqrt{1 - \left(\frac{B_0}{4c}\right)^2}} \right\}. \quad (5.37)$$

$$\mathcal{P}_{01}(z)=\mathcal{P}_{10}^{*}(z)= -\frac{j}{2}e^{-(B_0/2)z}\frac{\sin\left(2c\sqrt{1-\left(\frac{B_0}{4c}\right)^2}\cdot z\right)}{\sqrt{1-\left(\frac{B_0}{4c}\right)^2}}. \quad (5.38)$$

For $[B_0/(4c)] > 1$, the trigonometric functions may be rewritten as hyperbolic functions. For the limiting case,

$$\mathcal{P}_0(z) = 1 - \mathcal{P}_1(z) = \frac{1}{2} + \frac{1}{2}e^{-2cz}[1+2cz], \qquad B_0 = 4c. \quad (5.39)$$

$$\mathcal{P}_{01}(z) = \mathcal{P}_{10}^{*}(z) = -jcze^{-2cz}, \qquad B_0 = 4c. \quad (5.40)$$

Equation (5.37) is plotted in Figure 5.2, which shows the power in the excited mode of Figure 5.1 vs. distance for various propagation parameter imperfections.

With imperfections absent ($B_0/c = 0$) we obtain the result for the ideal coupler, given in Equation (5.5). As B_0/c increases the behavior changes from oscillatory to monotonic. For large z, the power divides equally between the two modes, except for the ideal case ($B_0 = 0$). Since the coupler has been assumed lossless, a plot for the power in the unexcited mode, $\mathcal{P}_1(z)$, is obtained as the mirror image of Figure 5.2 about the ordinate 0.5. $\mathcal{P}_0(z)$ and $\mathcal{P}_1(z)$, the total average powers in the two coupled guides, contain both coherent and incoherent components; the coherent components are the square magnitude of the results in Equation (5.29):

$$\begin{bmatrix}|\mathcal{I}_0(z)|^2\\|\mathcal{I}_1(z)|^2\end{bmatrix} = e^{-B_0 z}\begin{bmatrix}\cos^2 cz\\\sin^2 cz\end{bmatrix}. \quad (5.41)$$

Thus, in Figure 5.2, $\mathcal{P}_0(z)$ is completely coherent at $z = 0$, completely incoherent as $z \to \infty$.

5.5. DISCUSSION

We have found the expected response and average powers for a symmetric, uniform coupler with white random propagation parameters and constant coupling. The results are exact, and provide a

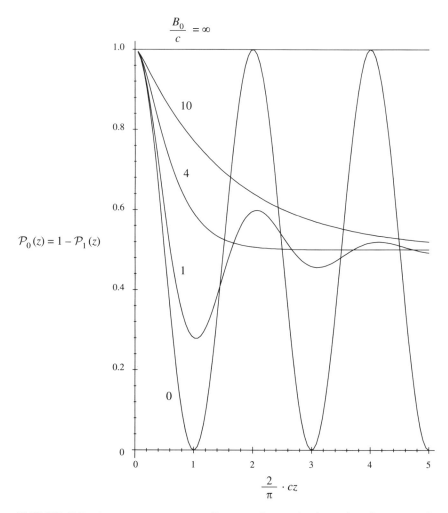

FIGURE 5.2. Average power vs. distance for excited mode of symmetric coupler.

good approximation for the almost white case. The analysis uses Kronecker product methods, similar to those of Chapter 3 for white coupling and constant propagation parameters. For two modes with white coupling, the average powers were decoupled from the cross-powers. Here the cross-powers are essential in the analysis.

These results are appropriate for investigating the necessary tolerances in two-branch couplers or switches, for which $B_0/c \ll 1$, and possible power dividers, where $B_0/c \approx 4$. Higher order statistics may be required to investigate the utility of the latter application.

The present methods may be applied to other related problems: correlated propagation parameters and coupling, asymmetric couplers (with different average propagation parameters for the two coupled guides), couplers with loss, multi-branch couplers, transfer-function and other higher order statistics, time-response statistics, etc.

REFERENCES

1. W. H. Louisell, *Coupled-Mode and Parametric Electronics*, Wiley, New York, 1960.
2. Theodor Tamir, Ed., *Guided-Wave Optoelectronics*, Springer-Verlag, New York, 1988.
3. Dietrich Marcuse, *Theory of Dielectric Optical Waveguides*, 2nd ed., Academic Press, New York, 1991.
4. Hermann A. Haus and Weiping Huang, "Coupled-Mode Theory," *Proceedings of the IEEE*, Vol. 79, October 1991, pp. 1505–1518.
5. Wei-Ping Huang, "Coupled-mode theory for optical waveguides: an overview," *J. Opt. Soc. Am. A*, Vol. 11, March 1994, pp. 963–983.
6. S. E. Miller, "Coupled Wave Theory and Waveguide Applications," *Bell System Technical Journal*, Vol. 33, May 1954, pp. 661–719.
7. Harrison E. Rowe and Iris M. Mack, "Coupled modes with random propagation constants," *Radio Science*, Vol. 16, July–August 1981, pp. 485–493.

CHAPTER SIX

Guides with General Coupling Spectra

6.1. INTRODUCTION

A variety of transmission statistics for two-mode and multi-mode guides having random coupling or propagation parameters with white spectra have been presented in Chapters 3–5. All of these results are exact; they provide a standard against which to compare approximate results for general imperfection spectra.

The matrix methods employing Kronecker products, used in earlier chapters, depend on dividing the guide into statistically independent sections of length Δz. For white coupling or propagation parameters, Δz can approach zero, different sections remain strictly independent, the analysis for each section becomes exact, and linear differential equations for the various transmission statistics follow directly from the matrix products. No perturbation or other approximations enter the analysis.

In order to apply Kronecker methods for general imperfection spectra, we again require the guide to be divided into independent sections. Now the section length Δz can no longer approach zero, but must be long compared to the correlation length of the random imperfections. The different sections are no longer strictly independent, but only approximately so. Exact analysis of the individual sections is no longer possible; we must now use perturbation theory for this purpose. Hence, the coupling is restricted to be small; the longer the correlation length, i.e., the narrower the bandwidth of the coupling or propagation parameter spectra, the smaller must be the coupling. The transmission statistics, expressed as matrix powers, are only approximate. Finally, differential equations for the approximate transmission statistics may not necessarily be obtained directly by setting $\Delta z \to 0$ in the matrix results.

We derive the coupled power equations for two-mode and N-mode guide with constant propagation parameters, for almost-white and for general non-white coupling spectra, in the present chapter. The range of validity of the present results is indicated. Other results of preceding chapters may be similarly generalized.

6.2. ALMOST-WHITE COUPLING SPECTRA

The covariance and spectrum of the coupling coefficient $c(z)$ are defined as the left-hand equalities of Equations (3.1) and (3.2):

$$R_c(\zeta) = \langle c(z+\zeta)c(z) \rangle. \tag{6.1}$$

$$S(\nu) = \int_{-\infty}^{\infty} R_c(\zeta) e^{-j2\pi\nu\zeta} d\zeta. \tag{6.2}$$

A typical low-pass coupling spectrum is shown in Figure 6.1.

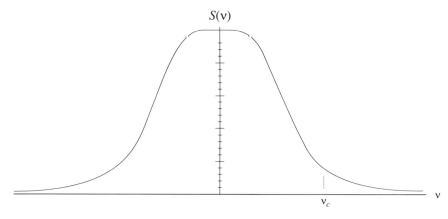

FIGURE 6.1. Low-pass coupling spectrum.

The cutoff spatial frequency ν_c gives a rough indication of the coupling bandwidth. The corresponding correlation distance is

$$\lambda_c = \frac{1}{\nu_c}. \tag{6.3}$$

The covariance will be small for $\zeta \gg \lambda_c$; consequently $c(z)$ and $c(z+\zeta)$ will be almost uncorrelated, and for Gaussian and many other $c(z)$ almost independent, for sufficiently large ζ. We assume that guide sections of length

$$\Delta z \gg \lambda_c \tag{6.4}$$

will be approximately independent, as required by the present Kronecker product methods.

In addition, we use the discrete approximation of Sections 2.4 and 2.6. The relative phase shift and relative attenuation between every pair of modes must therefore be small in length dz.

6.2.1. Two Modes

From Equations (2.30) and (6.3) and (6.4) [1],

$$\frac{1}{\nu_c} \ll \Delta z \ll \frac{1}{\Delta\alpha}, \quad \frac{1}{\nu_c} \ll \Delta z \ll \frac{2\pi}{|\Delta\beta|}. \tag{6.5}$$

Therefore,

$$\nu_c \gg \Delta\alpha, \quad \nu_c \gg \frac{|\Delta\beta|}{2\pi}. \tag{6.6}$$

A physical description of the almost-white case may be given in two equivalent ways in the lossless case: 1. $c(z)$ must vary rapidly in a beat wavelength $2\pi/|\Delta\beta|$. 2. The relative phase shift must be small in a correlation interval λ_c. These assumptions were previously used in Equations (4.65)–(4.68).

Equation (3.27) for the average transfer function, and the average of Equations (3.48) and (3.49) for the average powers and crosspowers, involve the statistical averages $\langle \sin c_k \rangle$, $\langle \cos c_k \rangle$, $\langle \sin^2 c_k \rangle$, $\langle \cos^2 c_k \rangle$, and $\langle \sin c_k \cos c_k \rangle$, where c_k is given by Equation (2.26). We assume c_k has a symmetric probability density, consistent with our earlier assumption that c_k has zero mean. Then,

$$\langle c_k \rangle = 0, \quad \langle \sin c_k \rangle = 0, \quad \langle \sin c_k \cos c_k \rangle = \tfrac{1}{2}\langle \sin 2c_k \rangle = 0. \tag{6.7}$$

Under the present restrictions of Equation (6.5), c_k will be approximately Gaussian in many cases. Then, extending the footnote to Equation (3.28),

$$\langle c_k^2 \rangle = \int_{(k-1)\Delta z}^{k\Delta z} \int_{(k-1)\Delta z}^{k\Delta z} R_c(z - z')dz\,dz'$$

$$\approx \int_{(k-1)\Delta z}^{k\Delta z} dz \int_{-\infty}^{\infty} R_c(\zeta)d\zeta = S(0)\Delta z. \quad (6.8)$$

For c_k approximately Gaussian,

$$\langle \cos c_k \rangle \approx e^{-\frac{1}{2}S(0)\Delta z}. \quad (6.9)$$

$$\langle \cos^2 c_k \rangle \approx \tfrac{1}{2}(1 + e^{-2S(0)\Delta z}); \quad \langle \sin^2 c_k \rangle \approx \tfrac{1}{2}(1 - e^{-2S(0)\Delta z}). \quad (6.10)$$

For simplicity, we restrict further analysis to the small coupling case. Expanding the trigonometric functions in Taylor series:

$$\langle \cos c_k \rangle \approx 1 - \tfrac{1}{2}\langle c_k^2 \rangle \approx 1 - \tfrac{1}{2}S(0)\Delta z; \quad S(0)\Delta z \ll 1. \quad (6.11)$$

$$\langle \cos^2 c_k \rangle \approx 1 - \langle c_k^2 \rangle \approx 1 - S(0)\Delta z,$$
$$\langle \sin^2 c_k \rangle \approx \langle c_k^2 \rangle \approx S(0)\Delta z; \quad S(0)\Delta z \ll 1. \quad (6.12)$$

Equation (6.5) yields the necessary condition

$$S(0)\frac{1}{\nu_c} = S(0)\lambda_c \ll 1. \quad (6.13)$$

The condition of Equation (6.13) is well satisfied for the numerical parameters of Equation (4.67).

Equations (6.5), (6.7), and (6.11) substituted in Equation (3.27) yield the following difference equations for the average transfer functions, where we have dropped explicit notation indicating that these results are approximate:

$$\mathcal{I}_0(k\Delta z) - \mathcal{I}_0[(k-1)\Delta z] = -\left[\Gamma_0 + \frac{S(0)}{2}\right]\mathcal{I}_0[(k-1)\Delta z]\Delta z;$$

$$\mathcal{I}_1(k\Delta z) - \mathcal{I}_1[(k-1)\Delta z] = -\left[\Gamma_1 + \frac{S(0)}{2}\right]\mathcal{I}_1[(k-1)\Delta z]\Delta z;$$

$$S(0)\Delta z \ll 1. \quad (6.14)$$

Equations (6.5), (6.7), and (6.12) substituted in the expected value of Equations (3.48) and (3.49) yield the following matrix difference equations for the average powers and cross-powers:

$$\begin{bmatrix} \mathcal{P}_0(k\Delta z) \\ \mathcal{P}_1(k\Delta z) \end{bmatrix} = \begin{bmatrix} 1 - [2\alpha_0 + S(0)]\Delta z & S(0)\Delta z \\ S(0)\Delta z & 1 - [2\alpha_1 + S(0)]\Delta z \end{bmatrix}$$
$$\cdot \begin{bmatrix} \mathcal{P}_0[(k-1)\Delta z] \\ \mathcal{P}_1[(k-1)\Delta z] \end{bmatrix}, \quad S(0)\Delta z \ll 1. \quad (6.15)$$

$$\begin{bmatrix} \mathcal{P}_{01}(k\Delta z) \\ \mathcal{P}_{10}(k\Delta z) \end{bmatrix} = \begin{bmatrix} 1 - [\Gamma_0 + \Gamma_1^* + S(0)]\Delta z & S(0)\Delta z \\ S(0)\Delta z & 1 - [\Gamma_0^* + \Gamma_1 + S(0)]\Delta z \end{bmatrix}$$
$$\cdot \begin{bmatrix} \mathcal{P}_{01}[(k-1)\Delta z] \\ \mathcal{P}_{10}[(k-1)\Delta z] \end{bmatrix}, \quad S(0)\Delta z \ll 1. \quad (6.16)$$

Δz is restricted by Equation (6.5) in the present almost-white case; it can no longer approach 0 as for strictly white coupling in Sections 3.3 and 3.4. Nevertheless, in the small coupling case Appendix I shows that useful approximations are obtained by taking the limit as $\Delta z \to 0$ of Equations (6.14)–(6.16), to yield the following differential equations.

Average transfer functions:

$$\mathcal{I}_0'(z) = -\left(\Gamma_0 + \frac{S(0)}{2}\right)\mathcal{I}_0(z),$$
$$\mathcal{I}_1'(z) = -\left(\Gamma_1 + \frac{S(0)}{2}\right)\mathcal{I}_1(z), \quad S(0)\Delta z \ll 1. \quad (6.17)$$

Coupled power equations:

$$\mathcal{P}_0'(z) = -[2\alpha_0 + S(0)]\mathcal{P}_0(z) + S(0)\mathcal{P}_1(z),$$
$$\mathcal{P}_1'(z) = S(0)\mathcal{P}_0(z) - [2\alpha_1 + S(0)]\mathcal{P}_1(z), \quad S(0)\Delta z \ll 1. \quad (6.18)$$

Cross-powers:

$$\mathcal{P}_{01}'(z) = -[\Gamma_0 + \Gamma_1^* + S(0)]\mathcal{P}_{01}(z) + S(0)\mathcal{P}_{10}(z),$$
$$\mathcal{P}_{10}'(z) = S(0)\mathcal{P}_{01}(z) - [\Gamma_0^* + \Gamma_1 + S(0)]\mathcal{P}_{10}(z), \quad S(0)\Delta z \ll 1.$$
$$(6.19)$$

Equations (6.17)–(6.19) for almost-white coupling are similar to the corresponding results of Equations (3.29), (3.53), and (3.54) for strictly white coupling with S_0 replaced by $S(0)$.

For large coupling, we must use Equations (6.9) and (6.10) in place of Equations (6.11) and (6.12) to obtain difference equations for the average transfer functions, average powers, and cross-powers. Equivalent differential equations are obtained by the more complicated procedure given in Appendix I, but differ from those of Equations (6.17)–(6.19).

6.2.2. N Modes

Equations (2.66) for constant propagation parameters, and Equations (6.3) and (6.4) yield

$$\frac{1}{v_c} \ll \Delta z \ll \frac{1}{|\Delta \alpha_{ij}|}, \qquad \frac{1}{v_c} \ll \Delta z \ll \frac{2\pi}{|\Delta \beta_{ij}|}; \qquad 0 \le i, j \le N-1. \tag{6.20}$$

We require the averages $\langle e^{jc_k \mathbf{C}} \rangle$ and $\langle e^{jc_k \mathbf{C}} \otimes e^{-jc_k \mathbf{C}} \rangle$ in Equations (3.37) and (3.58)–(3.59), where the coupling coefficient matrix $\mathbf{C}$ is given by Equation (3.35). Toward this end, diagonalize $\mathbf{C}$ as follows:

$$\mathbf{C} = \mathbf{L} \cdot \mathbf{\Lambda} \cdot \mathbf{L}^{-1}, \tag{6.21}$$

where

$$\mathbf{\Lambda} = \begin{bmatrix} \lambda_0 & & & 0 \\ & \lambda_1 & & \\ & & \ddots & \\ 0 & & & \lambda_{N-1} \end{bmatrix}. \tag{6.22}$$

Then,

$$c_k \mathbf{C} = \mathbf{L} \cdot c_k \mathbf{\Lambda} \cdot \mathbf{L}^{-1}. \tag{6.23}$$

$$e^{jc_k \mathbf{C}} = \mathbf{L} \cdot \begin{bmatrix} e^{jc_k \lambda_0} & & & 0 \\ & e^{jc_k \lambda_1} & & \\ & & \ddots & \\ 0 & & & e^{jc_k \lambda_{N-1}} \end{bmatrix} \cdot \mathbf{L}^{-1}. \qquad (6.24)$$

Under the assumptions of the previous section, c_k is approximately Gaussian with zero mean; therefore from Equation (6.8)

$$\langle e^{jc_k \mathbf{C}} \rangle = \mathbf{L} \cdot \begin{bmatrix} e^{-\frac{1}{2}\lambda_0^2 S(0)\Delta z} & & & 0 \\ & e^{-\frac{1}{2}\lambda_1^2 S(0)\Delta z} & & \\ & & \ddots & \\ 0 & & & e^{-\frac{1}{2}\lambda_{N-1}^2 S(0)\Delta z} \end{bmatrix} \cdot \mathbf{L}^{-1}. \qquad (6.25)$$

For small coupling, we may expand the exponentials to obtain

$$\langle e^{jc_k \mathbf{C}} \rangle$$

$$= \mathbf{L} \cdot \begin{bmatrix} 1 - \frac{1}{2}\lambda_0^2 S(0)\Delta z & & & 0 \\ & 1 - \frac{1}{2}\lambda_1^2 S(0)\Delta z & & \\ & & \ddots & \\ 0 & & & 1 - \frac{1}{2}\lambda_{N-1}^2 S(0)\Delta z \end{bmatrix} \cdot \mathbf{L}^{-1}$$

$$= \mathcal{I} - \frac{1}{2} S(0)\Delta z \mathbf{L} \cdot \begin{bmatrix} \lambda_0^2 & & & 0 \\ & \lambda_1^2 & & \\ & & \ddots & \\ 0 & & & \lambda_{N-1}^2 \end{bmatrix} \cdot \mathbf{L}^{-1}$$

$$= \mathcal{I} - \frac{1}{2} S(0)\Delta z \mathbf{C}^2, \qquad S(0)\Delta z \lambda_i^2 \big|_{\max} \ll 1, \qquad (6.26)$$

where $\mathcal{I}$ is the $N \times N$ unit matrix. Alternatively, we may use the left-hand equality of Equation (3.40) with Equation (6.8) to obtain this small-coupling result directly. Equation (6.26) may be substituted into Equation (3.37), and the limit taken as $\Delta z \to 0$ (Appendix I), to yield the result of Equation (3.41) with S_0 replaced by $S(0)$.

Utilizing the results of Appendix C, Equations (6.8) and (6.24) yield

$$\langle e^{jc_k \mathbf{C}} \otimes e^{-jc_k \mathbf{C}} \rangle = (\mathbf{L} \otimes \mathbf{L}) \cdot \mathbf{D} \cdot (\mathbf{L} \otimes \mathbf{L})^{-1}, \quad (6.27)$$

where $\mathbf{D}$ is a diagonal matrix, with elements

$$e^{-\frac{1}{2}(\lambda_i - \lambda_j)^2 S(0) \Delta z}, \quad 0 \leq i, j \leq N - 1. \quad (6.28)$$

For small coupling, we may substitute Equation (6.8) into Equation (3.60) and take the limit as in the preceding paragraph, obtaining finally the result of Equation (3.61) with $S_0 \to S(0)$.

To summarize, for a low-pass coupling spectrum satisfying the conditions of Equation (6.20) and the small-coupling condition of the final line of Equation (6.26), the average transfer functions and coupled power equations are obtained by the substitution $S_0 \to S(0)$ in the corresponding results for white coupling. This holds true also for the approximate results of Section 4.6 for the nondegenerate case, i.e., for distinct $\Delta \beta_{ij}$. Large coupling requires the use of Equations (6.25) and (6.27)–(6.28), and the extended procedure of Appendix I.

6.3. GENERAL COUPLING SPECTRA—LOSSLESS CASE

We now consider general coupling spectra with small coupling. Representative spectra are shown in Figures 6.1 and 6.2 for low-pass and band-pass coupling, respectively.

The guide is divided in to sections of length Δz long compared to the coupling correlation length λ_c, Equation (6.4). The guide transmission properties are described by perturbation theory, Sections 2.5 and 2.6 for two-modes and N-modes, respectively.

For two modes,

$$\begin{bmatrix} G_0(z) \\ G_1(z) \end{bmatrix} = \begin{bmatrix} 1 + H_0 & jH_{01} \\ jH_{10} & 1 + H_1 \end{bmatrix} \cdot \begin{bmatrix} G_0(z - \Delta z) \\ G_1(z - \Delta z) \end{bmatrix}, \quad (6.29)$$

6.3. GENERAL COUPLING SPECTRA—LOSSLESS CASE

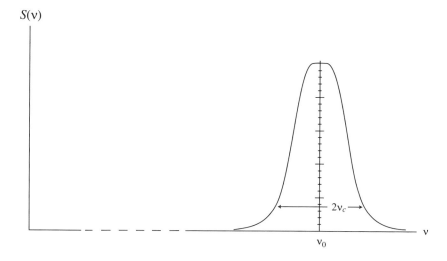

FIGURE 6.2. Band-pass coupling spectrum.

$$\begin{bmatrix} I_0(z) \\ I_1(z) \end{bmatrix} = \begin{bmatrix} e^{-\Gamma_0 \Delta z}(1 + H_0) & e^{-\Gamma_1 \Delta z} e^{-\Delta \Gamma z} jH_{01} \\ e^{-\Gamma_0 \Delta z} e^{+\Delta \Gamma z} jH_{10} & e^{-\Gamma_1 \Delta z}(1 + H_1) \end{bmatrix} \cdot \begin{bmatrix} I_0(z - \Delta z) \\ I_1(z - \Delta z) \end{bmatrix}, \tag{6.30}$$

where

$$\begin{aligned} H_0 &= -\int_0^{\Delta z} e^{+\Delta \Gamma \zeta} d\zeta \int_{z - \Delta z}^{z - \zeta} c(x)c(x + \zeta) dx, \\ H_{01} &= \int_{z - \Delta z}^{z} c(x) e^{+\Delta \Gamma x} dx, \\ H_{10} &= \int_{z - \Delta z}^{z} c(x) e^{-\Delta \Gamma x} dx, \\ H_1 &= -\int_0^{\Delta z} e^{-\Delta \Gamma \zeta} d\zeta \int_{z - \Delta z}^{z - \zeta} c(x)c(x + \zeta) dx. \end{aligned} \tag{6.31}$$

Small coupling requires

$$\int_{z-\Delta z}^{z} |c(x)| dx \ll 1. \tag{6.32}$$

This suggests the condition

$$\sqrt{\int_{-\infty}^{\infty} S(\nu) d\nu} \, \Delta z \ll 1. \tag{6.33}$$

For N modes,

$$I(z) = e^{-\Gamma z} \cdot G(z), \tag{6.34}$$

$$G(z) = \mathbf{G} \cdot G(z - \Delta z), \tag{6.35}$$

$$I(z) = e^{-\Gamma z} \cdot \mathbf{G} \cdot e^{+\Gamma(z - \Delta z)} \cdot I(z - \Delta z), \tag{6.36}$$

where

$$\mathbf{G} = \mathcal{I} + j \int_{z-\Delta z}^{z} c(x) e^{\Gamma x} \cdot \mathbf{C} \cdot e^{-\Gamma x} dx$$
$$- \int_{z-\Delta z}^{z} c(x) e^{\Gamma x} \cdot \mathbf{C} \cdot e^{-\Gamma x} dx \int_{z-\Delta z}^{x} c(y) e^{\Gamma y} \cdot \mathbf{C} \cdot e^{-\Gamma y} dy. \tag{6.37}$$

For the off-diagonal terms,

$$\mathbf{G}_{ij} = jC_{i-1,j-1} \int_{z-\Delta z}^{z} c(x) e^{\Delta \Gamma_{i-1,j-1} x} dx - \sum_{k=1}^{N} C_{i-1,k-1} C_{k-1,j-1}$$
$$\cdot \int_{0}^{\Delta z} e^{\Delta \Gamma_{i-1,k-1} \zeta} d\zeta \int_{z-\Delta z}^{z-\zeta} c(x) c(x+\zeta) e^{\Delta \Gamma_{i-1,j-1} x} dx,$$
$$1 \le i, j \le N, \ i \ne j. \tag{6.38}$$

The diagonal terms are

$$\mathbf{G}_{ii} = 1 - \sum_{k=1}^{N} C_{i-1,k-1} C_{k-1,i-1} \int_{0}^{\Delta z} e^{\Delta \Gamma_{i-1,k-1} \zeta} d\zeta$$
$$\cdot \int_{z-\Delta z}^{z-\zeta} c(x) c(x+\zeta) dx, \quad 1 \le i \le N. \tag{6.39}$$

The small coupling condition becomes

$$\int_{z-\Delta z}^{z} |c(x)| dx \|\mathbf{C}\| \ll 1, \tag{6.40}$$

where $\|\mathbf{C}\|$ denotes the matrix norm, Equation (A.25). The restriction corresponding to Equation (6.33) becomes

$$\sqrt{\int_{-\infty}^{\infty} S(\nu) d\nu} \Delta z \|\mathbf{C}\| \ll 1. \tag{6.41}$$

6.3. GENERAL COUPLING SPECTRA—LOSSLESS CASE

We consider the lossless case,

$$\alpha_{ij} = 0, \quad 0 \leq i, j \leq N-1, \tag{6.42}$$

throughout the rest of the present section.

6.3.1. Two Modes

The average transfer functions are found from the expected values of Equations (6.30)–(6.31). Substituting Equations (6.1)–(6.2) and (6.7), in the lossless case

$$\mathcal{I}_0(z) = e^{-j\beta_0 \Delta z}\left[1 - \Delta z \int_0^{\Delta z} R_c(\zeta)\left(1 - \frac{\zeta}{\Delta z}\right)e^{+j\Delta\beta\zeta}d\zeta\right]\mathcal{I}_0(z - \Delta z), \tag{6.43}$$

$$\mathcal{I}_1(z) = e^{-j\beta_1 \Delta z}\left[1 - \Delta z \int_0^{\Delta z} R_c(\zeta)\left(1 - \frac{\zeta}{\Delta z}\right)e^{-j\Delta\beta\zeta}d\zeta\right]\mathcal{I}_1(z - \Delta z), \tag{6.44}$$

where we use the notation of Equation (3.26) for the average complex mode amplitudes, $\mathcal{I}_i(z) = \langle I_i(z)\rangle$. By Equation (6.4), $R_c(\zeta)$ is narrow compared to $[1 - (\zeta/\Delta z)]$, and we have approximately

$$\mathcal{I}_0(z) = e^{-j\beta_0 \Delta z}\left\{1 - \frac{1}{2}\left[S\left(\frac{\Delta\beta}{2\pi}\right) + j\widehat{S}\left(\frac{\Delta\beta}{2\pi}\right)\right]\Delta z\right\}\mathcal{I}_0(z - \Delta z), \tag{6.45}$$

$$\mathcal{I}_1(z) = e^{-j\beta_1 z}\left\{1 - \frac{1}{2}\left[S\left(\frac{\Delta\beta}{2\pi}\right) - j\widehat{S}\left(\frac{\Delta\beta}{2\pi}\right)\right]\Delta z\right\}\mathcal{I}_1(z - \Delta z). \tag{6.46}$$

$\widehat{S}(\nu)$ is the Hilbert transform of the coupling spectrum $S(\nu)$ of Equation (6.2),

$$\widehat{S}(\nu) = \frac{1}{\pi}S(\nu) \star \frac{1}{\nu} = \frac{1}{\pi}\int_{-\infty}^{\infty}\frac{s(\mu)}{\nu - \mu}d\mu, \tag{6.47}$$

where $\star$ represents the convolution operator. Then, by the approximation of Appendix I,

$$\mathcal{I}_0(z) = e^{-j\beta_0 z}e^{-\frac{1}{2}z\left[S\left(\frac{\Delta\beta}{2\pi}\right) + j\widehat{S}\left(\frac{\Delta\beta}{2\pi}\right)\right]}\mathcal{I}_0(0), \tag{6.48}$$

$$\mathcal{I}_1(z) = e^{-j\beta_1 z} e^{-\frac{1}{2}z\left[S\left(\frac{\Delta\beta}{2\pi}\right) - j\hat{S}\left(\frac{\Delta\beta}{2\pi}\right)\right]} \mathcal{I}_1(0). \tag{6.49}$$

The coupled power equations are found by taking the Kronecker product of Equation (6.30) with its complex conjugate, using Equation (6.31), and taking the expected value of the result. Recalling the notation of Equation (3.50), we obtain the following approximation in the lossless case:

$$\begin{bmatrix} \mathcal{P}_0(z) \\ \mathcal{P}_{01}(z) \\ \mathcal{P}_{10}(z) \\ \mathcal{P}_1(z) \end{bmatrix} = \begin{bmatrix} B_{11} & 0 & 0 & B_{14} \\ 0 & B_{22} & B_{23} & 0 \\ 0 & B_{32} & B_{33} & 0 \\ B_{41} & 0 & 0 & B_{44} \end{bmatrix} \cdot \begin{bmatrix} \mathcal{P}_0(z - \Delta z) \\ \mathcal{P}_{01}(z - \Delta z) \\ \mathcal{P}_{10}(z - \Delta z) \\ \mathcal{P}_1(z - \Delta z) \end{bmatrix}, \tag{6.50}$$

where

$$B_{11} = B_{44} = 1 - \Delta z \int_{-\Delta z}^{\Delta z} R_c(\zeta) \left(1 - \frac{|\zeta|}{\Delta z}\right) e^{+j\Delta\beta\zeta} d\zeta,$$
$$B_{14} = B_{41} = \Delta z \int_{-\Delta z}^{\Delta z} R_c(\zeta) \left(1 - \frac{|\zeta|}{\Delta z}\right) e^{+j\Delta\beta\zeta} d\zeta. \tag{6.51}$$

The average powers $\mathcal{P}_0, \mathcal{P}_1$ are uncoupled from the cross-powers $\mathcal{P}_{01}, \mathcal{P}_{10}$; we confine our current interest to the average powers. Equation (6.4) permits neglect of the factor $[1 - (|\zeta|/\Delta z)]$ in Equation (6.51), and the approximation of Appendix I yields the coupled power equations for two modes:

$$\begin{bmatrix} \mathcal{P}_0'(z) \\ \mathcal{P}_1'(z) \end{bmatrix} = \begin{bmatrix} -S\left(\frac{\Delta\beta}{2\pi}\right) & S\left(\frac{\Delta\beta}{2\pi}\right) \\ S\left(\frac{\Delta\beta}{2\pi}\right) & -S\left(\frac{\Delta\beta}{2\pi}\right) \end{bmatrix} \cdot \begin{bmatrix} \mathcal{P}_0(z) \\ \mathcal{P}_1(z) \end{bmatrix}. \tag{6.52}$$

We omit the corresponding treatment of cross-powers.

6.3.2. N Modes

We retain the assumption of Equation (6.4) requiring guide sections to be approximately independent:

$$\Delta z \gg \lambda_c. \tag{6.53}$$

6.3. GENERAL COUPLING SPECTRA—LOSSLESS CASE

We seek coupled power equations for the non-degenerate case of Section 4.6, for large $|\Delta\beta_{ij}|$. For this purpose, we require the additional assumption

$$\Delta z \gg \frac{2\pi}{|\Delta\beta_{ij}|}, \qquad 0 \le i, j \le N-1; \tag{6.54}$$

i.e., the guide sections contain many beat wavelengths.

For the average transfer functions, take the expected value of Equation (6.35):

$$\mathcal{G}(z) = \langle G(z) \rangle = \langle \mathbf{G} \rangle \cdot \mathcal{G}(z - \Delta z), \tag{6.55}$$

where $\langle \mathbf{G} \rangle$ is the expected value of Equation (6.37), with elements obtained as the expected values of Equations (6.38) and (6.39). Using Equations (6.1)–(6.2) and (6.7), we have the following in the lossless case:

$$\langle \mathbf{G} \rangle_{ij} = -\sum_{k=1}^{N} C_{i-1,k-1} C_{k-1,j-1}$$

$$\cdot \Delta z \int_0^{\Delta z} R_c(\zeta) \left(e^{j\Delta\beta_{i-1,j-1}z} \frac{e^{-j\Delta\beta_{i-1,j-1}\zeta} - e^{-j\Delta\beta_{i-1,j-1}\Delta z}}{j\Delta\beta_{i-1,j-1}\Delta z} \right)$$

$$\cdot e^{j\Delta\beta_{i-1,k-1}\zeta} d\zeta, \qquad 1 \le i, j \le N, \ i \ne j. \tag{6.56}$$

$$\langle \mathbf{G} \rangle_{ii} = 1 - \sum_{k=1}^{N} C_{i-1,k-1} C_{k-1,i-1} \Delta z \int_0^{\Delta z} R_c(\zeta) \left(1 - \frac{\zeta}{\Delta z}\right) e^{j\Delta\beta_{i-1,k-1}\zeta} d\zeta,$$

$$1 \le i \le N. \tag{6.57}$$

The off-diagonal terms of Equation (6.56) may be neglected by Equation (6.54). The factor $[(1 - (\zeta/\Delta z)]$ in Equation (6.57) may be neglected by Equation (6.53), as in the corresponding derivation of Equations (6.45) and (6.46) for two modes. Therefore the following approximations hold in the non-degenerate case, indicated by the subscript ∞ as in Section 4.6:

$$\langle \mathbf{G} \rangle_{ii\infty} = 1 - \sum_{k=1}^{N} C_{i-1,k-1}^2 \frac{1}{2} \left[S\left(\frac{\Delta\beta_{i-1,k-1}}{2\pi}\right) + j\widehat{S}\left(\frac{\Delta\beta_{i-1,k-1}}{2\pi}\right) \right] \Delta z.$$

$$\tag{6.58}$$

$$\langle \mathbf{G} \rangle_{ij\infty} = 0, \quad i \neq j. \tag{6.59}$$

The expected value of Equation (6.35) yields

$$\mathcal{G}(z)_{\infty} = \langle \mathbf{G} \rangle_{\infty} \cdot \mathcal{G}(z - \Delta z)_{\infty}. \tag{6.60}$$

Changing notation,

$$\langle \mathbf{G} \rangle_{\infty} = \mathcal{I} - \mathbf{H}\Delta z, \tag{6.61}$$

where $\mathbf{H}$ is a diagonal matrix

$$\mathbf{H} = \begin{bmatrix} H_1 & & & 0 \\ & H_2 & & \\ & & \ddots & \\ 0 & & & H_N \end{bmatrix}, \tag{6.62}$$

with elements

$$H_i = \sum_{k=1}^{N} C_{i-1,k-1}^2 \frac{1}{2}\left[S\left(\frac{\Delta \beta_{i-1,k-1}}{2\pi}\right) + j\widehat{S}\left(\frac{\Delta \beta_{i-1,k-1}}{2\pi}\right) \right]. \tag{6.63}$$

From Appendix I, the corresponding approximate differential equation is

$$\mathcal{G}'(z)_{\infty} = -\mathbf{H} \cdot \mathcal{G}(z)_{\infty}, \tag{6.64}$$

and from Equation (6.34)

$$\mathcal{I}'(z)_{\infty} = -(j\boldsymbol{\beta} + \mathbf{H}) \cdot \mathcal{I}(z)_{\infty}. \tag{6.65}$$

$\boldsymbol{\beta}$ is the propagation matrix of Equation (2.58) in the lossless case. The expected complex amplitudes of the individual modes are

$$\mathcal{I}_i(z)_{\infty} = \langle I_i(z) \rangle_{\infty} = \langle I_i(0) \rangle e^{-j\beta_i z}$$

$$\cdot \exp\left\{ -\frac{1}{2} z \sum_{k=0}^{N-1} C_{ik}^2 \left[S\left(\frac{\Delta \beta_{ik}}{2\pi}\right) + j\widehat{S}\left(\frac{\Delta \beta_{ik}}{2\pi}\right) \right] \right\},$$

$$0 \leq i \leq N - 1. \tag{6.66}$$

6.3. GENERAL COUPLING SPECTRA—LOSSLESS CASE

The coupled power equations are obtained as the Kronecker product of Equation (6.34) with its complex conjugate. We seek the asymptotic results for large $|\Delta\beta_{ij}|$, Equation (6.54). We use the notation of Section 4.6.2. From Equations (4.91) and (4.101),

$$\mathcal{P}(z) = \left(e^{-\Gamma z} \otimes e^{-\Gamma^* z}\right) \cdot \mathcal{R}(z). \tag{6.67}$$

From Equation (4.95),

$$\mathcal{R}(z) = \langle G(z) \otimes G^*(z) \rangle. \tag{6.68}$$

Substituting Equation (6.35),

$$\mathcal{R}(z) = \mathbf{M} \cdot \mathcal{R}(z - \Delta z), \tag{6.69}$$

where

$$\mathbf{M} = \langle \mathbf{G} \otimes \mathbf{G}^* \rangle \tag{6.70}$$

and $\mathbf{G}$ is given by Equations (6.37)–(6.39). We now partition the matrices $\mathcal{R}(z)$ by Equation (4.106), and $\mathbf{M}$ as in Equation (4.108):

$$\mathbf{M} = \begin{bmatrix} \mathbf{M}_{|11} & \mathbf{M}_{|12} & \mathbf{M}_{|13} & \cdots & \mathbf{M}_{|1N} \\ \mathbf{M}_{|21} & \mathbf{M}_{|22} & \mathbf{M}_{|23} & \cdots & \mathbf{M}_{|2N} \\ \mathbf{M}_{|31} & \mathbf{M}_{|32} & \mathbf{M}_{|33} & \cdots & \mathbf{M}_{|3N} \\ \vdots & \vdots & \vdots & \ddots & \vdots \\ \mathbf{M}_{|N1} & \mathbf{M}_{|N2} & \mathbf{M}_{|N3} & \cdots & \mathbf{M}_{|NN} \end{bmatrix}. \tag{6.71}$$

The submatrices $\mathbf{M}_{|ij}$ are given by[1]

$$\mathbf{M}_{|ij} = (\mathbf{G})_{ij} \mathbf{G}^*. \tag{6.72}$$

The subvectors of $\mathcal{R}(z)$, defined by Equation (4.107), are obtained from Equations (6.69) and (6.71):

$$\mathcal{R}_i(z) = \sum_{j=1}^{N} \mathbf{M}_{|ij} \cdot \mathcal{R}_j(z - \Delta z). \tag{6.73}$$

---

[1]$(\mathbf{G})_{ij}$ represents the *ij* element of the matrix $\mathbf{G}$, previously denoted by the slightly different notation of Equation (6.38).

We seek the asymptotic properties of the $\mathbf{M}_{ij}$ as in Section 4.6.2, using similar calculations to those of Equations (6.56) and (6.57).

Consider the expected value of the off-diagonal submatrix $\mathbf{M}_{|ij}$; its $k\ell$ element is

$$\langle (\mathbf{M}_{|ij})_{k\ell} \rangle = \langle (\mathbf{G})_{ij}(\mathbf{G})^*_{k\ell} \rangle, \quad i \neq j. \tag{6.74}$$

All of the elements are small except for $k = i$, $\ell = j$; we have approximately

$$\langle (\mathbf{M}_{|ij})_{k\ell} \rangle_\infty = 0, \quad i \neq j, \quad k \neq i \text{ and/or } \ell \neq j. \tag{6.75}$$

$$\langle (\mathbf{M}_{|ij})_{ij} \rangle_\infty = C^2_{i-1,j-1} S\left(\frac{\Delta \beta_{i-1,j-1}}{2\pi} \right) \Delta z, \quad i \neq j. \tag{6.76}$$

Consider next the expected value of the diagonal submatrix $\mathbf{M}_{|ii}$; its $k\ell$ element is

$$\langle (\mathbf{M}_{|ii})_{k\ell} \rangle = \langle (\mathbf{G})_{ii}(\mathbf{G})^*_{k\ell} \rangle. \tag{6.77}$$

The off-diagonal elements are small, and neglected in the nondegenerate approximation:

$$\langle (\mathbf{M}_{|ii})_{k\ell} \rangle_\infty = 0, \quad k \neq \ell. \tag{6.78}$$

For the diagonal elements,

$$\langle (\mathbf{M}_{|ii})_{\ell\ell} \rangle_\infty = 1 - \sum_{k=1}^{N} C^2_{i-1,k-1} \frac{1}{2} \left[S\left(\frac{\Delta \beta_{i-1,k-1}}{2\pi} \right) + j\widehat{S}\left(\frac{\Delta \beta_{i-1,k-1}}{2\pi} \right) \right] \Delta z$$
$$- \sum_{k=1}^{N} C^2_{\ell-1,k-1} \frac{1}{2} \left[S\left(\frac{\Delta \beta_{\ell-1,k-1}}{2\pi} \right) - j\widehat{S}\left(\frac{\Delta \beta_{\ell-1,k-1}}{2\pi} \right) \right] \Delta z.$$
$$\tag{6.79}$$

The special case $\ell = i$ will be the only one of interest in determining the average powers:

$$\langle (\mathbf{M}_{|ii})_{ii} \rangle_\infty = 1 - \sum_{k=1}^{N} C^2_{i-1,k-1} S\left(\frac{\Delta \beta_{i-1,k-1}}{2\pi} \right) \Delta z. \tag{6.80}$$

Consider the ith component of the subvector $\mathcal{R}_i(z)$ of Equation (6.73). Recalling the notation of Equation (4.107),

$$\mathcal{R}_{i-1}(z)_\infty = \langle (\mathbf{M}_{|ii})_{ii} \rangle_\infty \mathcal{R}_{i-1}(z - \Delta z)_\infty + \sum_{\substack{j=1 \\ j \neq i}}^{N} \langle (\mathbf{M}_{|ij})_{ij} \rangle_\infty \mathcal{R}_{j-1}(z - \Delta z)_\infty. \quad (6.81)$$

Substituting Equations (6.76) and (6.80),

$$\mathcal{R}_{i-1}(z)_\infty = \left[1 - \sum_{k=1}^{N} C_{i-1,k-1}^2 S\left(\frac{\Delta \beta_{i-1,k-1}}{2\pi}\right) \Delta z \right] \mathcal{R}_{i-1}(z - \Delta z)_\infty$$

$$+ \left[\sum_{j=1}^{N} C_{i-1,j-1}^2 S\left(\frac{\Delta \beta_{i-1,j-1}}{2\pi}\right) \Delta z\right] \mathcal{R}_{j-1}(z - \Delta z)_\infty. \quad (6.82)$$

Relabeling the summation indices,

$$\mathcal{P}_i(z)_\infty = \mathcal{P}_i(z - \Delta z)_\infty - \sum_{k=0}^{N-1} C_{ik}^2 S\left(\frac{\Delta \beta_{ik}}{2\pi}\right) \mathcal{P}_i(z - \Delta z)_\infty \Delta z$$

$$+ \sum_{k=0}^{N-1} C_{ik}^2 S\left(\frac{\Delta \beta_{ik}}{2\pi}\right) \mathcal{P}_k(z - \Delta z)_\infty \Delta z. \quad (6.83)$$

Finally, from Appendix I the coupled power equations are given as follows in the nondegenerate case [2]:

$$\mathcal{P}_i'(z)_\infty = -\sum_{k=0}^{N-1} C_{ik}^2 S\left(\frac{\Delta \beta_{ik}}{2\pi}\right) \mathcal{P}_i(z)_\infty$$

$$+ \sum_{k=0}^{N-1} C_{ik}^2 S\left(\frac{\Delta \beta_{ik}}{2\pi}\right) \mathcal{P}_k(z)_\infty, \quad 0 \leq i \leq N-1. \quad (6.84)$$

6.4. GENERAL COUPLING SPECTRA—LOSSY CASE

Finally, we consider general coupling spectra for lossy media with small coupling. Equations (6.4) and (6.29)–(6.41) remain valid, but Equation (6.42) no longer holds.

116 GUIDES WITH GENERAL COUPLING SPECTRA

6.4.1. Two Modes

The expected values of Equations (6.30)–(6.31) yield the following generalizations of the lossless results, Equations (6.48)–(6.49):

$$\mathcal{I}_0(z) = e^{-\Gamma_0 z} e^{-\frac{1}{2}z\left[S_0\left(\frac{\Delta\beta}{2\pi}\right) + j\hat{S}_0\left(\frac{\Delta\beta}{2\pi}\right)\right]} \mathcal{I}_0(0), \tag{6.85}$$

$$\mathcal{I}_1(z) = e^{-\Gamma_1 z} e^{-\frac{1}{2}z\left[S_1\left(\frac{\Delta\beta}{2\pi}\right) - j\hat{S}_1\left(\frac{\Delta\beta}{2\pi}\right)\right]} \mathcal{I}_1(0), \tag{6.86}$$

where

$$S_0(\nu) = \int_{-\infty}^{\infty} R_c(\zeta) e^{-|\Delta\alpha||\zeta|} e^{-j2\pi\nu\zeta} d\zeta, \tag{6.87}$$

$$S_1(\nu) = \int_{-\infty}^{\infty} R_c(\zeta) e^{+|\Delta\alpha||\zeta|} e^{+j2\pi\nu\zeta} d\zeta, \tag{6.88}$$

and we assume that the signal mode has lower loss, i.e., $\Delta\alpha = \alpha_0 - \alpha_1 < 0$. Note that $S_0(\nu) \neq S_1(\nu)$, in contrast to the lossless case. $R_c(\zeta)$ must decay more rapidly than $e^{+|\Delta\alpha||\zeta|}$ grows for large $|\zeta|$, in order for Equation (6.88) to hold.

For the lossy coupled power equations, Equation (6.51) generalize as follows. For the diagonal elements,

$$B_{11} = e^{-2\alpha_0 \Delta z}\left[1 - \Delta z \int_{-\Delta z}^{\Delta z} R_c(\zeta) e^{-|\Delta\alpha||\zeta|}\left(1 - \frac{|\zeta|}{\Delta z}\right) e^{+j\Delta\beta\zeta} d\zeta\right],$$

$$B_{44} = e^{-2\alpha_1 \Delta z}\left[1 - \Delta z \int_{-\Delta z}^{\Delta z} R_c(\zeta) e^{+|\Delta\alpha||\zeta|}\left(1 - \frac{|\zeta|}{\Delta z}\right) e^{+j\Delta\beta\zeta} d\zeta\right]. \tag{6.89}$$

Since $R_c(\zeta)$ is narrow by Equation (6.4), we have the usual approximations, neglecting the factor $[1 - (|\zeta|/\Delta z)]$ in the integrands of Equation (6.89):

$$B_{11} = e^{-2\alpha_0 \Delta z}\left[1 - S_0\left(\frac{\Delta\beta}{2\pi}\right)\Delta z\right],$$

$$B_{44} = e^{-2\alpha_1 \Delta z}\left[1 - S_1\left(\frac{\Delta\beta}{2\pi}\right)\Delta z\right], \tag{6.90}$$

where $S_0(\ )$ and $S_1(\ )$ are given by Equations (6.87) and (6.88). Again, the product $R_c(\zeta) e^{+|\Delta\alpha||\zeta|}$ must decay for large $|\zeta|$ in order for this result for B_{44} to remain valid.

6.4. GENERAL COUPLING SPECTRA—LOSSY CASE

For the off-diagonal elements,

$$B_{14} = e^{-2\alpha_1 \Delta z} \Delta z \int_{-\Delta z}^{\Delta z} R_c(\zeta) e^{-\Delta \alpha |\zeta|} e^{j\Delta\beta\zeta} \frac{1 - e^{-2\Delta\alpha\Delta z\left(1 - \frac{|\zeta|}{\Delta z}\right)}}{2\Delta\alpha\Delta z} d\zeta,$$

$$B_{41} = e^{-2\alpha_0 \Delta z} \Delta z \int_{-\Delta z}^{\Delta z} R_c(\zeta) e^{+\Delta \alpha |\zeta|} e^{j\Delta\beta\cdot\zeta} \frac{1 - e^{+2\Delta\alpha\Delta z\left(1 - \frac{|\zeta|}{\Delta z}\right)}}{-2\Delta\alpha\Delta z} d\zeta.$$
(6.91)

These quantities can be expressed simply only for small differential loss in the section length Δz. In the low-loss case,

$$\frac{1 - e^{-2\Delta\alpha\Delta z\left(1 - \frac{|\zeta|}{\Delta z}\right)}}{2\Delta\alpha\Delta z} \approx \frac{1 - e^{+2\Delta\alpha\Delta z\left(1 - \frac{|\zeta|}{\Delta z}\right)}}{-2\Delta\alpha\Delta z}$$

$$\approx \left(1 - \frac{|\zeta|}{\Delta z}\right), \quad |\Delta\alpha|\Delta z \ll 1. \quad (6.92)$$

Then the same approximations used in deriving Equation (6.90) yield

$$B_{14} = e^{-2\alpha_1 \Delta z} S_1\left(\frac{\Delta\beta}{2\pi}\right) \Delta z,$$
$$B_{41} = e^{-2\alpha_0 \Delta z} S_0\left(\frac{\Delta\beta}{2\pi}\right) \Delta z, \quad |\Delta\alpha|\Delta z \ll 1. \quad (6.93)$$

Combining these results,

$$\begin{bmatrix} P_0(z) \\ P_1(z) \end{bmatrix} = e^{-(\alpha_0 + \alpha_1)\Delta z}$$

$$\cdot \begin{bmatrix} e^{+|\Delta\alpha|\Delta z}\left[1 - S_0\left(\frac{\Delta\beta}{2\pi}\right)\Delta z\right] & e^{-|\Delta\alpha|\Delta z} S_1\left(\frac{\Delta\beta}{2\pi}\right)\Delta z \\ e^{+|\Delta\alpha|\Delta z} S_0\left(\frac{\Delta\beta}{2\pi}\right)\Delta z & e^{-|\Delta\alpha|\Delta z}\left[1 - S_1\left(\frac{\Delta\beta}{2\pi}\right)\Delta z\right] \end{bmatrix}$$

$$\cdot \begin{bmatrix} P_0(z - \Delta z) \\ P_1(z - \Delta z) \end{bmatrix}, \quad |\Delta\alpha|\Delta z \ll 1. \quad (6.94)$$

118 GUIDES WITH GENERAL COUPLING SPECTRA

From the perturbation results of Appendix I, we obtain the coupled power equations for small differential loss:

$$\begin{bmatrix} P_0'(z) \\ P_1'(z) \end{bmatrix} = \begin{bmatrix} -2\alpha_0 - S_0\left(\dfrac{\Delta\beta}{2\pi}\right) & S_1\left(\dfrac{\Delta\beta}{2\pi}\right) \\ S_0\left(\dfrac{\Delta\beta}{2\pi}\right) & -2\alpha_1 - S_1\left(\dfrac{\Delta\beta}{2\pi}\right) \end{bmatrix}$$
$$\cdot \begin{bmatrix} P_0(z) \\ P_1(z) \end{bmatrix}, \quad |\Delta\alpha|\Delta z \ll 1. \quad (6.95)$$

Since we have assumed $\lambda_c \ll \Delta z$ in Equation (6.4), the exponential factors in Equations (6.87) and (6.88) will be insignificant, and $S_0(\ )$ and $S_1(\ )$ in Equation (6.95) differ only trivially from $S(\ )$ of Equation (6.2).

For significant differential loss, the extended procedure of Appendix I must be applied to Equations (6.89) and (6.91) to obtain coupled power equations.

6.4.2. N Modes

The lossy two-mode results of Section 6.4.1 are readily extended to N modes in the nondegenerate case. The lossless result of Equation (6.66) for the average transfer functions becomes

$$\mathcal{T}_i(z)_\infty = \langle I_i(z)\rangle_\infty = \langle I_i(0)\rangle e^{-\Gamma_i z}$$
$$\cdot \exp\left\{-\frac{1}{2}z \sum_{k=0}^{N-1} C_{ik}^2 \left[S_{ik}\left(\frac{\Delta\beta_{ik}}{2\pi}\right) + j\widehat{S_{ik}}\left(\frac{\Delta\beta_{ik}}{2\pi}\right)\right]\right\},$$
$$0 \leq i \leq N-1, \quad (6.96)$$

where

$$S_{ik}(\nu) = \int_{-\infty}^{\infty} R_c(\zeta) e^{\Delta\alpha_{ik}|\zeta|} e^{j2\pi\nu\zeta} d\zeta. \quad (6.97)$$

The lossless coupled power equations of Section 6.3.2 may be similarly extended to the lossy case. The quantity of Equation (6.76)

6.4. GENERAL COUPLING SPECTRA—LOSSY CASE

becomes

$$\langle(\mathbf{M}_{|ij})_{ij}\rangle_\infty = C^2_{i-1,j-1} \cdot e^{2\Delta\alpha_{i-1,j-1}z}\Delta z$$

$$\cdot \int_{-\Delta z}^{\Delta z} R_c(\zeta) e^{-\Delta\alpha_{i-1,j-1}|\zeta|} e^{j\Delta\beta_{i-1,j-1}\zeta} \frac{1 - e^{-2\Delta\alpha_{i-1,j-1}\Delta z\left(1-\frac{|\zeta|}{2\pi}\right)}}{2\Delta\alpha_{i-1,j-1}\Delta z} d\zeta,$$

$$i \neq j. \quad (6.98)$$

For small differential loss in each section, the approximation of Equation (6.92) yields

$$\langle(\mathbf{M}_{|ij})_{ij}\rangle_\infty = C^2_{i-1,j-1} \cdot e^{2\Delta\alpha_{i-1,j-1}z} S_{j-1,i-1}\left(\frac{\Delta\beta_{i-1,j-1}}{2\pi}\right)\Delta z,$$

$$|\Delta\alpha_{i-1,j-1}|\Delta z \ll 1. \quad (6.99)$$

However, the assumption of Equation (6.4) renders the exponential factor in Equation (6.97) unimportant, and $S_{j-1,i-1}(\ )$ in Equation (6.99) is well approximated by $S(\ )$ of Equation (6.2) in the low-loss case.

The quantity of Equation (6.80) becomes

$$\langle(\mathbf{M}_{|ii})_{ii}\rangle_\infty = 1 - \Delta z \sum_{k=1}^{N} C^2_{i-1,k-1}$$

$$\cdot \int_{-\Delta z}^{\Delta z} R_c(\zeta) e^{\Delta\alpha_{i-1,k-1}|\zeta|}\left(1 - \frac{|\zeta|}{\Delta z}\right) e^{j\Delta\beta_{i-1,k-1}\zeta} d\zeta. \quad (6.100)$$

Neglecting the quantity $[1-(|\zeta|/\Delta z)]$ by the assumption of Equation (6.4), the usual approximations yield

$$\langle(\mathbf{M}_{|ii})_{ii}\rangle_\infty = 1 - \Delta z \sum_{k=1}^{N} C^2_{i-1,k-1} S_{i-1,k-1}\left(\frac{\Delta\beta_{i-1,k-1}}{2\pi}\right). \quad (6.101)$$

These results do *not* require small loss; for the low-loss case $S_{i-1,k-1}(\ )$ is again approximated by $S(\ )$ of Equation (6.2).

Substituting Equations (6.98) and (6.101) into Equation (6.81), the lossless results of Equations (6.82) and (6.83) are modified as follows for the lossy case:

$$\mathcal{R}_{i-1}(z)_\infty = \left[1 - \sum_{k=1}^{N} C_{i-1,k-1}^2 S_{i-1,k-1}\left(\frac{\Delta\beta_{i-1,k-1}}{2\pi}\right)\Delta z\right]\mathcal{R}_{i-1}(z-\Delta z)_\infty$$

$$+ \left[\sum_{j=1}^{N} C_{i-1,j-1}^2 e^{2\Delta\alpha_{i-1,j-1}z}\Delta z\right.$$

$$\cdot \int_{-\Delta z}^{\Delta z} R_c(\zeta)e^{-\Delta\alpha_{i-1,j-1}|\zeta|}e^{j\Delta\beta_{i-1,j-1}\zeta}$$

$$\left. \cdot \frac{1 - e^{-2\Delta\alpha_{i-1,j-1}\Delta z\left(1-\frac{|\zeta|}{2\pi}\right)}}{2\Delta\alpha_{i-1,j-1}\Delta z} d\zeta \mathcal{R}_{j-1}(z-\Delta z)_\infty\right]. \quad (6.102)$$

$$\mathcal{P}_i(z)_\infty = \left[1 - \sum_{k=0}^{N-1} C_{ik}^2 S_{ik}\left(\frac{\Delta\beta_{ik}}{2\pi}\right)\Delta z\right]e^{-2\alpha_i\Delta z}\mathcal{P}_i(z-\Delta z)_\infty$$

$$+ \left[\sum_{k=0}^{N-1} C_{ik}^2 \Delta z \int_{-\Delta z}^{\Delta z} R_c(\zeta)e^{-\Delta\alpha_{ik}|\zeta|}e^{j\Delta\beta_{ik}\zeta}\right.$$

$$\left. \cdot \frac{1 - e^{-2\Delta\alpha_{ik}\Delta z\left(1-\frac{|\zeta|}{2\pi}\right)}}{2\Delta\alpha_{ik}\Delta z} d\zeta\, e^{-2\alpha_k\Delta z}\mathcal{P}_k(z-\Delta z)_\infty\right]. \quad (6.103)$$

For small differential loss in each section, Equation (6.99) replaces Equation (6.98), and Equation (6.103) becomes

$$\mathcal{P}_i(z)_\infty = \left[1 - \sum_{k=0}^{N-1} C_{ik}^2 S\left(\frac{\Delta\beta_{ik}}{2\pi}\right)\Delta z\right]e^{-2\alpha_i\Delta z}\mathcal{P}_i(z-\Delta z)_\infty$$

$$+ \sum_{k=0}^{N-1} C_{ik}^2 \Delta z \cdot S\left(\frac{\Delta\beta_{ik}}{2\pi}\right)e^{-2\alpha_k\Delta z}\mathcal{P}_k(z-\Delta z)_\infty,$$

$$|\Delta\alpha_{ik}|\Delta z \ll 1, \quad 0 \leq i, k \leq N - 1. \quad (6.104)$$

We have used the approximation that $S_{ik}(\ )$ of Equation (6.97) is approximately equal to $S(\ )$ of Equation (6.2) in the low-loss case. The perturbation results of Appendix I yield the coupled power equations for small loss per section [2]:

$$\mathcal{P}'_i(z)_\infty = -\left[2\alpha_i + \sum_{k=0}^{N-1} C_{ik}^2 S\left(\frac{\Delta\beta_{ik}}{2\pi}\right)\right]\mathcal{P}_i(z)_\infty$$

$$+ \sum_{k=0}^{N-1} C_{ik}^2 \cdot S\left(\frac{\Delta\beta_{ik}}{2\pi}\right)\mathcal{P}_k(z)_\infty,$$

$$|\Delta\alpha_{ik}|\Delta z \ll 1, \quad 0 \le i, k \le N-1. \quad (6.105)$$

The extended procedure of Appendix I must be applied to Equation (6.102) or (6.103) for significant differential loss.

6.5. DISCUSSION

Exact solutions for the transmission statistics of multi-mode guides have been presented for white coupling in Chapters 3 and 4. These results depend on the single parameter S_0, the coupling spectral density.

This chapter extends these results to general coupling spectra $S(\nu)$, for average transfer functions and coupled power equations of two-mode and nondegenerate N-mode guides. This generalization is accomplished by combining the earlier Kronecker product methods with perturbation theory. Other transmission statistics treated in earlier chapters may be similarly extended. These results are no longer exact, but require the coupling to be small.[2] However, the small-coupling restrictions of Equations (6.33) and (6.41) are only heuristic.

For the lossless case the present results replace S_0 of the white-coupling results by the general coupling spectral density $S[(\Delta\beta_{ij}/2\pi)]$, calculated at appropriate values of its argument. However, a similar result is true for the coupled power equations in the lossy case only for low differential loss; otherwise the extended procedure of Appendix I is required.

---

[2] An exception occurs for the almost-white case of Section 6.2, which permits the treatment of large coupling by the extended procedure of Appendix I. The limiting case, completely degenerate guides, is treated in Appendix G.

122 GUIDES WITH GENERAL COUPLING SPECTRA

The necessity of this more general procedure is illustrated by considering the term $S_1[\Delta\beta/(2\pi)]$ in the matrix of Equation (6.95) for the covariance

$$R_c(\zeta) = 1 - |\zeta|, \qquad |\zeta| \leq 1, \qquad (6.106)$$

with differential attenuation

$$|\Delta\alpha| = 1. \qquad (6.107)$$

Substituting these relations into Equation (6.88),

$$S_1(.884) = -0.163. \qquad (6.108)$$

Negative values clearly violate the physical requirement that powers must be positive for any input conditions [3].

REFERENCES

1. Harrison E. Rowe, "Waves with Random Coupling and Random Propagation Constants," *Applied Scientific Research*, Vol. 41, 1984, pp. 237–255.
2. Dietrich Marcuse, *Theory of Dielectric Optical Waveguides*, 2nd ed., Academic Press, New York, 1991.
3. Neil A. Jackman, *Limitations and Tolerances in Optical Devices*, Appendix C, Ph.D. Thesis, Stevens Institute of Technology, Castle Point, Hoboken, NJ, 1994.

CHAPTER SEVEN

Four-Mode Guide with Exponential Coupling Covariance

7.1. INTRODUCTION

We illustrate the results of Chapter 6 for general coupling spectra with a single example:

1. Exponential coupling covariance.
2. Four modes.
3. Nondegenerate.
4. Lossless.

Other examples appear in Chapter 8.

The covariance and spectrum of the coupling coefficient, Equations (6.1) and (6.2), are given as follows:

$$R_c(\zeta) = R_c(0)e^{-|\zeta|/\zeta_c}. \tag{7.1}$$

$$S(\nu) = R_c(0)\frac{2/\zeta_c}{(1/\zeta_c)^2 + (2\pi\nu)^2}. \tag{7.2}$$

The Hilbert transform of the coupling spectrum, Equation (6.47), is

$$\widehat{S}(\nu) = R_c(0)\frac{4\pi\nu}{(1/\zeta_c)^2 + (2\pi\nu)^2}. \tag{7.3}$$

The following numerical parameters apply to the four-mode dielectric slab waveguide of Appendix F.2. From Equations (F.20) and (F.21),

$$\Gamma = j\boldsymbol{\beta} = j \times 10^6 \begin{bmatrix} 6.3424 & 0 & 0 & 0 \\ 0 & 6.3316 & 0 & 0 \\ 0 & 0 & 6.3142 & 0 \\ 0 & 0 & 0 & 6.2921 \end{bmatrix} \text{ m}^{-1}. \quad (7.4)$$

$$\mathbf{C} = 10^9 \begin{bmatrix} 0 & 2.3225 & 0 & 4.4966 \\ 2.3225 & 0 & 6.8737 & 0 \\ 0 & 6.8737 & 0 & 13.3083 \\ 4.4966 & 0 & 13.3083 & 0 \end{bmatrix} \text{ m}^{-2}. \quad (7.5)$$

We obtain the following matrix of differential propagation constants from Equation (7.4):

$$\Delta\boldsymbol{\beta} = \begin{bmatrix} 0 & \Delta\beta_{01} & \Delta\beta_{02} & \Delta\beta_{03} \\ \Delta\beta_{10} & 0 & \Delta\beta_{12} & \Delta\beta_{13} \\ \Delta\beta_{20} & \Delta\beta_{21} & 0 & \Delta\beta_{23} \\ \Delta\beta_{30} & \Delta\beta_{31} & \Delta\beta_{32} & 0 \end{bmatrix}$$

$$= 10^6 \begin{bmatrix} 0 & 0.0108 & 0.0282 & 0.0503 \\ -0.0108 & 0 & 0.0174 & 0.0395 \\ -0.0282 & -0.0174 & 0 & 0.0221 \\ -0.0503 & -0.0395 & -0.0221 & 0 \end{bmatrix} \text{ m}^{-1}. \quad (7.6)$$

We next select numerical values for the parameters $R_c(0)$ and ζ_c of Equations (7.1) and (7.2). We take

$$\zeta_c = 8 \times 10^{-5} \text{ m}. \quad (7.7)$$

The cutoff frequency and wavelength of $S(\nu)$ of Equation (7.2) are then

$$\nu_c = \frac{1}{2\pi\zeta_c} = 1.9894 \times 10^3 \text{ m}^{-1},$$

$$\lambda_c = 2\pi\zeta_c = 5.0265 \times 10^{-4} \text{ m}. \quad (7.8)$$

7.1. INTRODUCTION

This choice places the coupling cutoff wavelength in the middle of the beat wavelengths of the four modes, which range from $2\pi/\Delta\beta_{03} = 1.250 \times 10^{-4}$ m. to $2\pi/\Delta\beta_{01} = 5.818 \times 10^{-4}$ m.

Δz must satisfy two constraints, Equations (6.53) and (6.54). We take

$$\Delta z = 0.005 \text{ m.} \tag{7.9}$$

The small-coupling condition of Equation (6.41) becomes

$$\sqrt{R_c(0)}\Delta z \|\mathbf{C}\| \ll 1. \tag{7.10}$$

From Equations (7.5) and (A.25),

$$\|\mathbf{C}\| = 2.0182 \times 10^{10} \text{ m}^{-2}. \tag{7.11}$$

Combining these parameters,

$$\sqrt{R_c(0)} \ll 9.91 \times 10^{-9} \text{ m.} \tag{7.12}$$

We take

$$\sqrt{R_c(0)} = 5 \times 10^{-10} \text{ m.}, \qquad R_c(0) = 2.5 \times 10^{-19} \text{ m}^2. \tag{7.13}$$

Substituting these numerical values, Equations (7.1) and (7.2) yield the following for the present example:

$$R_c(\zeta) = 2.5 \times 10^{-19} e^{-1.25 \times 10^{+4}|\zeta|}. \tag{7.14}$$

$$S(\nu) = \frac{6.25 \times 10^{-15}}{1.5625 \times 10^{+8} + (2\pi\nu)^2}. \tag{7.15}$$

$$\widehat{S}(\nu) = \frac{\pi\nu \times 10^{-18}}{1.5625 \times 10^{+8} + (2\pi\nu)^2}. \tag{7.16}$$

$S(0) = 4 \times 10^{-23}$. The present low-pass coupling spectrum has much smaller bandwidth, low-frequency spectral density, and rms coupling and straightness deviation than the almost-white example of Equations (4.67) and (4.68).

7.2. AVERAGE TRANSFER FUNCTIONS

Assume deterministic inputs $I_i(0)$, $0 \leq i \leq 3$. Take the magnitude of Equation (6.66), to yield

$$|\mathcal{I}_i(z)|_\infty = I_i(0) \exp\left[-\frac{1}{2}z \sum_{k=0}^{3} C_{ik}^2 S\left(\frac{\Delta \beta_{ik}}{2\pi}\right)\right], \quad 0 \leq i \leq 3. \quad (7.17)$$

The expected response magnitudes are found by substituting the numerical parameters of Section 7.1 into this relation:

$$\begin{aligned}
|\mathcal{I}_0(z)|_\infty &= I_0(0) \exp(-8.5295 \times 10^{-5} z). \\
|\mathcal{I}_1(z)|_\infty &= I_1(0) \exp(-3.8328 \times 10^{-4} z). \\
|\mathcal{I}_2(z)|_\infty &= I_2(0) \exp(-1.1792 \times 10^{-3} z). \\
|\mathcal{I}_3(z)|_\infty &= I_3(0) \exp(-8.8120 \times 10^{-4} z).
\end{aligned} \quad (7.18)$$

Finally, the angle of Equation (6.66) for real (zero phase) inputs is

$$\angle \mathcal{I}_i(z)_\infty = -z\left[\beta_i + \frac{1}{2} \sum_{k=0}^{3} C_{ik}^2 \widehat{S}\left(\frac{\Delta \beta_{ik}}{2\pi}\right)\right]. \quad (7.19)$$

For the numerical parameters of Section 7.1, the expected response angles are

$$\begin{aligned}
\angle \mathcal{I}_0(z)_\infty &= -z(\beta_0 + 1.480 \times 10^{-4}), \\
\angle \mathcal{I}_1(z)_\infty &= -z(\beta_1 + 3.943 \times 10^{-4}), \\
\angle \mathcal{I}_2(z)_\infty &= -z(\beta_2 + 1.070 \times 10^{-3}), \\
\angle \mathcal{I}_3(z)_\infty &= -z(\beta_3 + 1.612 \times 10^{-3}).
\end{aligned} \quad (7.20)$$

The propagation constants β_i, given in Equation (7.4), are large compared to the contributions of the random coupling.

7.3. COUPLED POWER EQUATIONS

We obtain the coupled power equations by substituting Equations (7.5), (7.6), and (7.15) into Equation (6.84), with $N = 4$:

$$\begin{bmatrix} \mathcal{P}'_0(z)_\infty \\ \mathcal{P}'_1(z)_\infty \\ \mathcal{P}'_2(z)_\infty \\ \mathcal{P}'_3(z)_\infty \end{bmatrix} = 10^{-4} \begin{bmatrix} -1.7059 & 1.2358 & 0 & 0.4701 \\ 1.2358 & -7.6655 & 6.4297 & 0 \\ 0 & 6.4297 & -23.5839 & 17.1542 \\ 0.4701 & 0 & 17.1542 & -17.6243 \end{bmatrix} \cdot \begin{bmatrix} \mathcal{P}_0(z)_\infty \\ \mathcal{P}_1(z)_\infty \\ \mathcal{P}_2(z)_\infty \\ \mathcal{P}_3(z)_\infty \end{bmatrix}. \quad (7.21)$$

7.4. DISCUSSION

The results of the present chapter illustrate the coupled power equations of Chapter 6 for the particular dielectric slab waveguide of Appendix F with straightness deviation spectrum of Equation (7.15). Other results of prior chapters for white imperfections may be similarly extended to general spectra.

Such results are only approximate; they require the imperfections to be small. For the present problem this requirement is given in Equation (7.12). However, such limits are only heuristic; no rigorous bounds on errors are available.

Exact transmission statistics for non-white coupling or propagation parameters are available only for random square-wave imperfections. Chapter 8 derives results for random square-wave coupling. Comparison with these exact results permits validation of the approximate transmission statistics of Chapter 6.

CHAPTER EIGHT

Random Square-Wave Coupling

8.1. INTRODUCTION

We derive exact transmission statistics for coupled modes with random square-wave coupling coefficients having a variety of statistics. Such transmission lines consist of cascaded sections, each having constant coupling, with transfer functions described by modification of Equations (2.13)–(2.18) and (2.59), for the two-mode and multimode cases, respectively.

Chapters 3 and 4 have presented exact transmission statistics for white coupling. However, general coupling spectra, treated in Chapter 6, permit only approximate treatment, with heuristic estimates for the range of validity. The present results yield exact transmission statistics for a variety of coupling spectra corresponding to random square-wave coupling models, and thus permit validation of these approximate results in the present special cases.

The present methods have additional interest, in that they extend the prior analysis, based on products of statistically independent matrices, to products of Markov matrices.

A square-wave coupling function is illustrated in Figure 8.1. The coupling $c(z)$ has constant value c_k in the kth section:

$$c(z) = c_k, \quad (k-1)\ell < z < k\ell. \tag{8.1}$$

We give exact coupled power equations for a number of statistical models for $\{c_k\}$, presented in Appendix J. The spectra for these

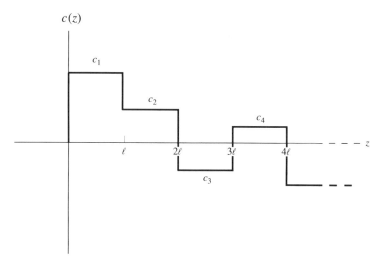

FIGURE 8.1. Square-wave coupling.

$c(z)$, also presented in this appendix, permit comparison of these exact results with the approximations of Chapter 6. Numerical results, obtained using MAPLE, are presented for the signal power $\mathcal{P}_0(z)$, for single-mode signal input with unit power:

$$\mathcal{P}_0(0) = \langle |I_0(0)|^2 \rangle = 1; \quad \mathcal{P}_i(0) = 0, \quad 1 \le i.$$
$$I_0(0) = 1, \quad \text{coherent input}; \quad \langle I_0(0) \rangle = 0, \quad \text{incoherent input}.$$
(8.2)

The coherent components (i.e., expected values) of all transfer functions are identically zero for incoherent input.

The transmission matrix characterizing the kth section of a lossless two-mode guide with square-wave coupling of Figure 8.1 is obtained from Equations (2.13)–(2.18) as follows:

$$\begin{bmatrix} I_0(k\ell) \\ I_1(k\ell) \end{bmatrix} = \mathbf{T}(c_k) \cdot \begin{bmatrix} I_0[(k-1)\ell] \\ I_1[(k-1)\ell] \end{bmatrix}. \tag{8.3}$$

$$\mathbf{T}(c_k) = e^{-j\frac{\beta_0+\beta_1}{2}\ell}/(K_+ - K_-)$$
$$\cdot \begin{bmatrix} -K_- e^{j(\Delta\beta/2)\ell}\sqrt{} + K_+ e^{-j(\Delta\beta/2)\ell}\sqrt{} & e^{j(\Delta\beta/2)\ell}\sqrt{} - e^{-j(\Delta\beta/2)\ell}\sqrt{} \\ e^{j(\Delta\beta/2)\ell}\sqrt{} - e^{-j(\Delta\beta/2)\ell}\sqrt{} & K_+ e^{j(\Delta\beta/2)\ell}\sqrt{} - K_- e^{-j(\Delta\beta/2)\ell}\sqrt{} \end{bmatrix}.$$
(8.4)

$$K_{\pm} = \frac{1 \pm \sqrt{}}{2c_k/\Delta\beta}; \quad K_+ K_- = -1. \tag{8.5}$$

$$\frac{1}{K_+ - K_-} = \frac{c_k/\Delta\beta}{\sqrt{}}. \tag{8.6}$$

$$\sqrt{} = \sqrt{1 + (2c_k/\Delta\beta)^2}. \tag{8.7}$$

$$\Delta\beta = \beta_0 - \beta_1. \tag{8.8}$$

The corresponding result for the lossless multi-mode case follows from Equations (2.43), (2.45), (2.58), and (2.59):

$$\mathbf{I}(k\ell) = \mathbf{T}(c_k) \cdot \mathbf{I}[(k-1)\ell]. \tag{8.9}$$

$$\mathbf{T}(c_k) = e^{j\{-\boldsymbol{\beta} + c_k \mathbf{C}\}z}. \tag{8.10}$$

The four-mode example of Appendix F illustrates a more realistic physical system; $\boldsymbol{\beta}$ and $\mathbf{C}$ are given by Equations (F.20) and (F.21):

$$\boldsymbol{\Gamma} = j\boldsymbol{\beta} = j \times 10^6 \begin{bmatrix} 6.3424 & 0 & 0 & 0 \\ 0 & 6.3316 & 0 & 0 \\ 0 & 0 & 6.3142 & 0 \\ 0 & 0 & 0 & 6.2921 \end{bmatrix} \text{m}^{-1}. \tag{8.11}$$

$$\mathbf{C} = 10^9 \begin{bmatrix} 0 & 2.3225 & 0 & 4.4966 \\ 2.3225 & 0 & 6.8737 & 0 \\ 0 & 6.8737 & 0 & 13.3083 \\ 4.4966 & 0 & 13.3083 & 0 \end{bmatrix} \text{m}^{-2}. \tag{8.12}$$

For 2 modes, Equations (8.9) and (8.10) specialize as follows:

$$\boldsymbol{\beta} = \begin{bmatrix} \beta_0 & 0 \\ 0 & \beta_1 \end{bmatrix}, \tag{8.13}$$

132 RANDOM SQUARE-WAVE COUPLING

and from Equation (2.60)

$$\mathbf{C} = \begin{bmatrix} 0 & 1 \\ 1 & 0 \end{bmatrix}. \tag{8.14}$$

These results are equivalent to Equations (8.3)–(8.8).

8.2. TWO MODES—BINARY INDEPENDENT SECTIONS

Equations (J.21) and (J.22) give the coupling statistics for independent, symmetric, binary coupling. The coupling spectrum is given by Equation (J.24), and plotted in Figure 8.2.

The small-coupling condition of Equations (6.32) and (6.33) becomes

$$a\ell \ll 1. \tag{8.15}$$

Equation (6.52) yields the following approximation for the signal power with input given by Equation (8.2):

$$P_{0app}(z) = \frac{1}{2} + \frac{1}{2} e^{-2S(\frac{\Delta\beta}{2\pi})z}. \tag{8.16}$$

For n sections, total guide length $z = n\ell$, Equations (8.3), (D.14), and (D.17) yield the following exact solutions for the inputs of Equation (8.2):

$$\begin{bmatrix} \mathcal{I}_0(n\ell) \\ \mathcal{I}_1(n\ell) \end{bmatrix} = \begin{bmatrix} \langle I_0(n\ell) \rangle \\ \langle I_1(n\ell) \rangle \end{bmatrix} = \langle \mathbf{T}(c_k) \rangle^n \cdot \begin{bmatrix} 1 \\ 0 \end{bmatrix}, \quad \text{coherent input.} \tag{8.17}$$

$$\begin{bmatrix} \mathcal{I}_0(n\ell) \\ \mathcal{I}_1(n\ell) \end{bmatrix} = \begin{bmatrix} \langle I_0(n\ell) \rangle \\ \langle I_1(n\ell) \rangle \end{bmatrix} = \begin{bmatrix} 0 \\ 0 \end{bmatrix}, \quad \text{incoherent input.} \tag{8.18}$$

$$\begin{bmatrix} \mathcal{P}_0(n\ell) \\ \mathcal{P}_{01}(n\ell) \\ \mathcal{P}_{10}(n\ell) \\ \mathcal{P}_1(n\ell) \end{bmatrix} = \langle \mathbf{T}(c_k) \otimes \mathbf{T}^*(c_k) \rangle^n \cdot \begin{bmatrix} 1 \\ 0 \\ 0 \\ 0 \end{bmatrix},$$

coherent or incoherent input, (8.19)

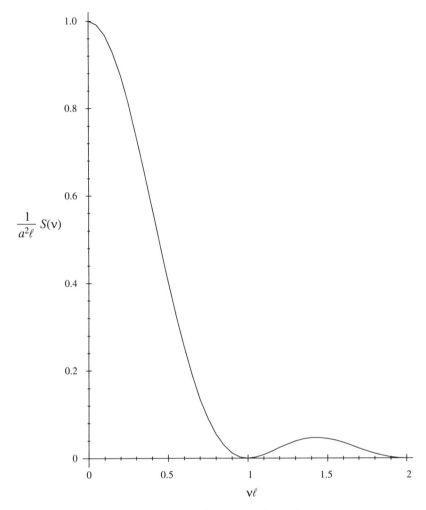

FIGURE 8.2. Coupling spectrum: binary independent square-wave coupling.

where the powers and cross-powers are defined in Equation (3.50). In these equations,

$$\langle \mathbf{T}(c_k) \rangle = \frac{1}{2}[\mathbf{T}(a) + \mathbf{T}(-a)], \qquad (8.20)$$

$$\langle \mathbf{T}(c_k) \otimes \mathbf{T}^*(c_k) \rangle = \frac{1}{2}[\mathbf{T}(a) \otimes \mathbf{T}^*(a) + \mathbf{T}(-a) \otimes \mathbf{T}^*(-a)], \qquad (8.21)$$

where $\mathbf{T}(\ )$ is given by Equations (8.4)–(8.8).

134 RANDOM SQUARE-WAVE COUPLING

$\mathcal{P}_{0app}$ of Equation (8.16), and the exact $\mathcal{P}_0$ determined from Equations (8.19) and (8.21) are plotted in Figure 8.3 for $a\ell = 0.5$ and $\Delta\beta\ell = 0.2\pi$; the exact plot consists of discrete points since Equation (8.19) contains an integral number of sections. The value of $a\ell$ for this example is less than 1, but is *not* much less than 1, as required by the heuristic condition of Equation (8.15). Therefore the approximate and exact plots in Figure 8.3 are close, but not identical. As $a\ell$ decreases these two results approach each other rapidly.

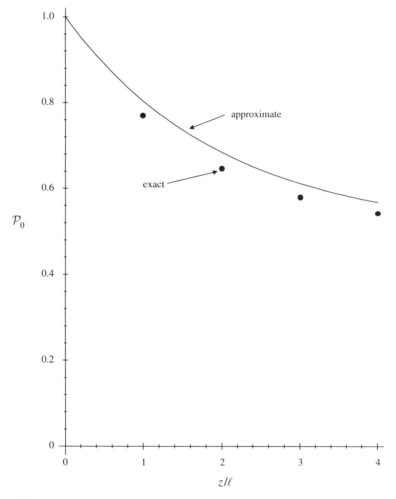

FIGURE 8.3. Signal power vs. normalized distance; two modes; binary independent square-wave coupling. $a\ell = 0.5$, $\Delta\beta\ell = 0.2\pi$.

8.3. TWO MODES—BINARY MARKOV SECTIONS

A great variety of coupling spectra are available in the Markov case; the coupling statistics are described in Section J.2.2. The parameter p, the probability that two adjacent c_k in Equation (8.1) and Figure 8.1 are different, determines the shape of the coupling spectrum. The transition matrix, unconditional probability vector, and coupling vector for the symmetric binary case are given in Equation (D.36). Equation (J.27) for the coupling spectrum $S(\nu)$ yields the result of Figure 8.2 for $p = \frac{1}{2}$, the independent case. For small p, the coupling approaches a constant and the spectrum approaches a δ function; as $p \to 1$ the coupling approaches a periodic square wave, and $S(\nu)$ approaches a line spectrum with fundamental frequency $\nu = 0.5/\ell$. Figures 8.4 and 8.5 show typical $S(\nu)$ for small and large p.

We choose the band-pass $S(\nu)$ of Figure 8.5 with $p = 0.8$ for the following example, with $\Delta\beta\ell = \pi$, corresponding to the point

$$\nu\ell = 0.5, \quad S(\nu) = 1.621 a^2 \ell, \tag{8.22}$$

close to the peak of $S(\nu)$. The small-coupling requirement remains as in Equation (8.15). The approximate signal power obtained by substituting the above numerical value in Equation (8.16) is

$$\mathcal{P}_{0app}(z) = \frac{1}{2} + \frac{1}{2} e^{-3.242 \cdot (a\ell)^2 \frac{z}{\ell}}. \tag{8.23}$$

The exact coupled power equations for unit signal input, Equation (8.2), are given from Equations (D.58) and (D.59):

$$\begin{bmatrix} \mathcal{P}_0(n\ell) \\ \mathcal{P}_{01}(n\ell) \\ \mathcal{P}_{10}(n\ell) \\ \mathcal{P}_1(n\ell) \end{bmatrix} = \frac{1}{2} [\mathcal{I} \; \mathcal{I}] \cdot \begin{bmatrix} 0.2 \mathbf{A}_1 \otimes \mathbf{A}_1^* & 0.8 \mathbf{A}_1 \otimes \mathbf{A}_1^* \\ 0.8 \mathbf{A}_2 \otimes \mathbf{A}_2^* & 0.2 \mathbf{A}_2 \otimes \mathbf{A}_2^* \end{bmatrix}^{(n-1)}$$

$$\cdot \begin{bmatrix} \mathbf{A}_1 \otimes \mathbf{A}_1^* \\ \mathbf{A}_2 \otimes \mathbf{A}_2^* \end{bmatrix} \cdot \begin{bmatrix} 1 \\ 0 \\ 0 \\ 0 \end{bmatrix}, \tag{8.24}$$

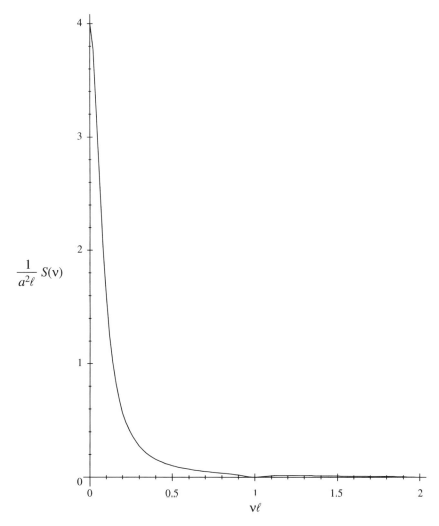

FIGURE 8.4. Coupling spectrum: binary Markov square-wave coupling. $p = 0.2$.

where $n = z/\ell$, $\mathcal{I}$ is the 4×4 unit matrix, and

$$\mathbf{A}_1 = \mathbf{T}(a), \quad \mathbf{A}_2 = \mathbf{T}(-a), \tag{8.25}$$

where $\mathbf{T}(\ )$ is given by Equations (8.4)–(8.8).

The approximate and exact signal power, $\mathcal{P}_{0app}$ of Equation (8.23) and $\mathcal{P}_0(n\ell)$ of Equation (8.24), are plotted in Figure 8.6 for $a\ell = 0.2$

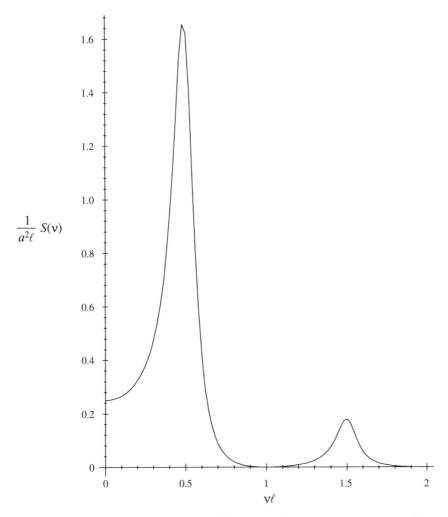

FIGURE 8.5. Coupling spectrum: binary Markov square-wave coupling. $p = 0.8$.

and $\Delta\beta\ell = \pi$ (as noted above, corresponding approximately to the peak of the band-pass coupling spectrum of Figure 8.5).

As in the preceding section, we have chosen a value of $a\ell$ somewhat larger than indicated by Equation (8.15), in order to display a significant difference between the approximate and exact solutions in Figure 8.6.

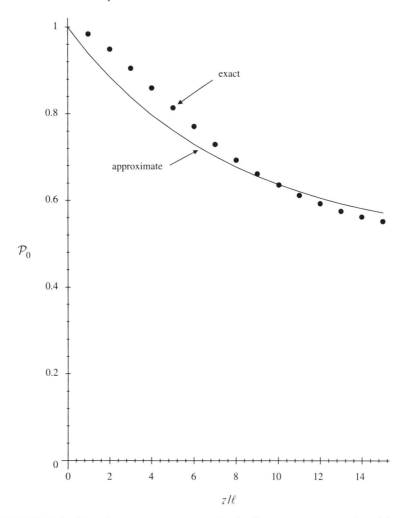

FIGURE 8.6. Signal power vs. normalized distance; two modes; binary Markov square-wave coupling. $a\ell = 0.2$, $\Delta\beta\ell = \pi$.

8.4. FOUR MODES—MULTI-LEVEL MARKOV SECTIONS

We take as a final example the two multi-level coupling coefficient models described by Equations (J.9)–(J.17) and Section J.3. The parameters c_k of Equation (8.1) and Figure 8.1 for these two models may be described as follows:

$$c_k = d\left[d_{k-2} + d_{k-1} + d_k + d_{k+1} + d_{k+2}\right], \quad \text{Low-Pass.} \quad (8.26)$$

8.4. FOUR MODES—MULTI-LEVEL MARKOV SECTIONS

$$c_k = d\left[d_{k-2} + d_{k-1} - d_{k+1} - d_{k+2}\right], \quad \text{Band-Pass.} \quad (8.27)$$

The separation between coupling levels is $2d$. The parameters $\{d_k\}$ are independent, symmetric, binary random variables taking on the values ± 1. The corresponding coupling spectra are given by Equations (J.29) and (J.46). Both generate pseudo-Gaussian square-wave random coupling functions, with six and five levels, respectively. The coupling spectra for these two cases are shown in Figures 8.7 and 8.8. We choose the low-pass case for the following numerical example; the corresponding probability distribution is given in Equation (J.30).

The larger number of parameters requires treatment of a specific numerical example, in contrast to the two-mode case. We take the four-mode guide of Appendix F, with parameters given in Equations (8.9)–(8.12). The mean-square coupling coefficient is found from either Equation (J.29) or (J.30) as

$$\langle c_k^2 \rangle = \int_{-\infty}^{\infty} S(\nu)d\nu = 5d^2. \quad (8.28)$$

The matrix norm is found from Equations (A.25) and (8.12) as

$$\|C\| = 20.1820 \times 10^9 \text{ m}^{-2}. \quad (8.29)$$

We choose

$$d = 10^{-6} \text{ m}, \quad (8.30)$$

$$\ell = 2 \times 10^{-5} \text{ m}. \quad (8.31)$$

Then the spectral width of $S(\nu)$ in Figure 8.7 is 10^4 m^{-1}, roughly comparable to that of the low-pass example in Chapter 4, in Equation (4.66). However, the low-pass spectral density, $S(0) = 5 \times 10^{-16}$ m^3, is much greater than that of Equation (4.67). The root mean square (rms) straightness deviation is

$$\sqrt{\langle c^2(z) \rangle} = \sqrt{\int_{-\infty}^{\infty} S(\nu)d\nu} = 2.236 \times 10^{-6} \text{ m}, \quad (8.32)$$

also much larger than that of Equation (4.68), but still smaller than the guide width of 1.23×10^{-5} m from Equation (F.17). The left-hand

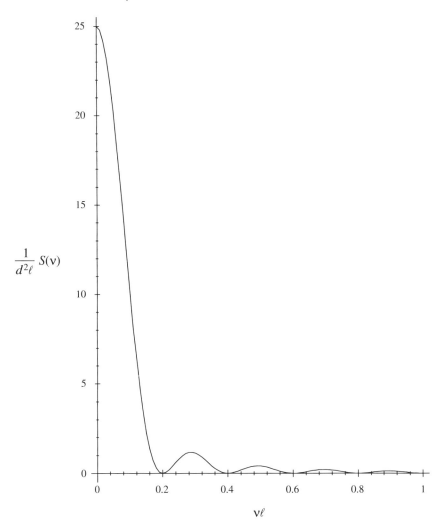

FIGURE 8.7. Coupling spectrum: six-level Markov square-wave coupling.

side of the small coupling condition of Equation (6.41) becomes

$$\sqrt{\int_{-\infty}^{\infty} S(v)dv\ell} \, \|C\| = 0.903. \tag{8.33}$$

This is *not* small compared to 1; we have made the coupling large for the present example.

The coherent signal component for the coherent input of Equation (8.2) is given by Equations (6.66), (J.29), (8.11), and (8.12) for

8.4. FOUR MODES—MULTI-LEVEL MARKOV SECTIONS

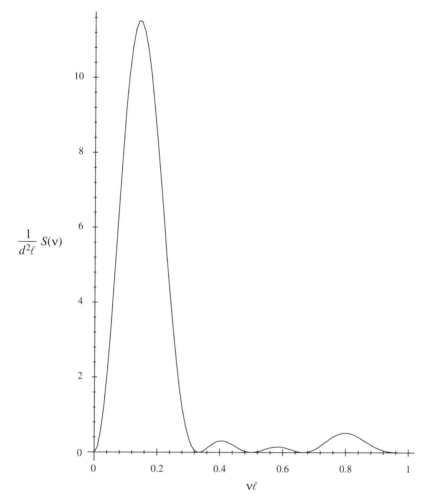

FIGURE 8.8. Coupling spectrum: five-level Markov square-wave coupling.

the small coupling approximation. Substituting the above numerical parameters, we have

$$|\langle \mathcal{I}_0(z) \rangle|_{app} = \exp\left\{-\frac{1}{2}z\left[C_{01}^2 S\left(\frac{\Delta\beta_{01}}{2\pi}\right) + C_{02}^2 S\left(\frac{\Delta\beta_{02}}{2\pi}\right)\right]\right\} \quad (8.34)$$
$$= e^{-1496z}.$$

The corresponding exact quantity $|\langle \mathcal{I}_0(z) \rangle|$ is found from Equations (D.50) and (D.39)–(D.43). The $\mathbf{A}_i$ of Equation (D.41) are given by

142 RANDOM SQUARE-WAVE COUPLING

Equation (8.10) as

$$\mathbf{A}_i = \mathbf{T}(a_i) = e^{j\{-\boldsymbol{\beta} + da_i \mathbf{C}\}\ell}, \tag{8.35}$$

where the propagation and coupling matrices $\boldsymbol{\beta}$ and $\mathbf{C}$ are given by Equations (8.11) and (8.12). These approximate and exact results are plotted in Figure 8.9.

The small-coupling condition of Equation (6.41) is significantly violated by Equation (8.33); the difference between the two results in Figure 8.9 is not surprising.

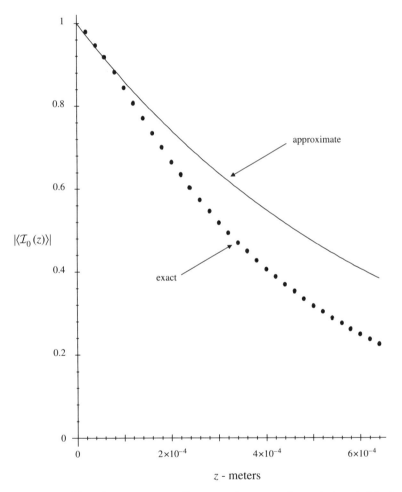

FIGURE 8.9. Signal power vs. distance; four modes; six-level Markov square-wave coupling.

8.5. DISCUSSION

Exact results for the coupled line equations with random coupling or propagation parameters are available for the limiting case of white imperfection spectra. General coupling spectra, treated in Chapter 6, require the use of both Kronecker product methods, to find the overall transmission statistics of cascaded sections, and perturbation theory, to determine the performance of a typical section. These results are therefore only approximate; heuristic conditions for small coupling are given in Equations (6.33) and (6.41) for two modes and N modes, respectively.

The present chapter has extended Kronecker product methods to Markov matrices, permitting the treatment of correlated sections. The choice of square-wave coupling permits exact treatment of the individual sections, and therefore yields exact transmission statistics for square-wave coupling with a variety of spectra. These results offer partial validation of the intuitive small-coupling conditions of Section 6.3.

The present results suggest the following physical interpretation of the small-coupling requirement: perturbation theory must apply in a correlation length of the coupling coefficient $c(z)$.

The final example, of Section 8.4, represents an approach to a more realistic model of a random coupling function. The example chosen requires numerical computing with quite large matrices—128×128 in these calculations for the expected transfer functions. These results are within the capabilities of MAPLE on a Pentium machine running at 233 MHz. Significant reduction in computing time is achieved by taking advantage of the special properties of the matrices in this particular example.

Higher order statistics require correspondingly increased computing resources. Determining the mode powers would require matrix products 512×512 in size.

Finally, our choice of square-wave coupling for the individual sections permits exact results. However, the Markov–Kronecker product methods may be applied to other coupling functions for the individual sections, greatly enlarging the scope of these methods. The individual sections would no longer permit exact treatment, as in the present case, but would have to be solved by perturbation theory or other numerical methods.

CHAPTER NINE

Multi-Layer Coatings with Random Optical Thickness[1]

9.1. INTRODUCTION

Figure 9.1 shows a multi-layer coating, designed to serve as an optical filter, mirror, or window [3–6]. It consists of ℓ dielectric layers with thickness d_k and index of refraction $n_k = \sqrt{\epsilon_k/\epsilon_0}$, $1 \le k \le \ell$. The medium to the left of the coating has index n_i, the output medium has index n_t; as a special case either or both may be free space, with $n = 1$.

An input plane wave with electric field E_i is incident upon the coating, producing a reflected wave with field E_r and an output wave with field E_t. The complex transfer functions for reflection and transmission are

$$R = \frac{E_r}{E_i}, \quad T = \frac{E_t}{E_i}. \tag{9.1}$$

The reflectance and transmittance are the corresponding ratios of intensities:

$$\mathcal{R} = |R|^2, \quad \mathcal{T} = |T|^2 \cdot \frac{n_t}{n_i}. \tag{9.2}$$

The design values for $\{d_k\}$ and $\{n_k\}$ are chosen to realize the desired transmission and reflection performance. We study the re-

---
[1] This chapter is based on thesis work of Neil Jackman [1, 2].

146 MULTI-LAYER COATINGS WITH RANDOM OPTICAL THICKNESS

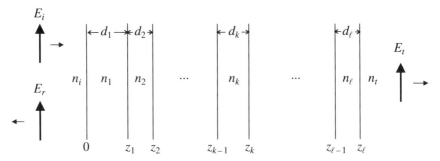

FIGURE 9.1. Multi-layer dielectric structure.

sulting transfer function statistics due to random variations in these parameters. The following assumptions are made:

1. Infinite parallel plane geometry.
2. Dielectric layers of uniform thickness and dielectric constant.
3. Input wave linearly polarized, at normal incidence.
4. Ideal design given.
5. Lossless dielectrics.

Since the layers are lossless,

$$\mathcal{R} + \mathcal{T} = 1. \tag{9.3}$$

The restriction to normal incidence is convenient for illustrating the present methods, but not essential.

Forward and backward plane waves are present at the input, and in each layer; the output is assumed matched, so only a forward wave is present there. These two waves are coupled by the dielectric layers; both can have significant magnitude. This problem therefore corresponds to the analysis of previous chapters with two coupled modes, but with a significant difference. Here the two modes travel in opposite directions; in former chapters all modes traveled in the same direction. The response of this system is given by Equation (K.18) as

$$\begin{bmatrix} E_i \\ E_r \end{bmatrix} = \frac{1}{2} \begin{bmatrix} 1 & \frac{1}{n_i} \\ 1 & -\frac{1}{n_i} \end{bmatrix} \cdot \prod_{k=1}^{\ell} \mathbf{M}_k \cdot \begin{bmatrix} 1 & 1 \\ n_t & -n_t \end{bmatrix} \cdot \begin{bmatrix} E_t \\ 0 \end{bmatrix}, \tag{9.4}$$

where $\mathbf{M}_k$ is given by Equations (K.15) and (K.16):

$$\mathbf{M}_k = \begin{bmatrix} \cos\phi_k & \dfrac{j\sin\phi_k}{n_k} \\ jn_k \sin\phi_k & \cos\phi_k \end{bmatrix}, \quad \phi_k = \frac{2\pi}{\lambda} n_k d_k. \qquad (9.5)$$

λ is the free-space wavelength; the n's and d's are shown in Figure 9.1.

The circuit-theoretic model of a dominant-mode transmission line with random parameters is closely related to the present problem [7]. In both, input and output are related by matrix products, permitting Kronecker product methods to be applied. However, unlike the forward-mode case of preceding chapters, the outputs are known while the inputs are not. Therefore direct analysis yields the statistics for only the *reciprocal* transfer function $1/T$. We desire statistics for T and for R, corresponding to mixed boundary conditions at input and output, i.e., a unit forward wave at the input and zero backward wave at the output.

A heuristic solution for the average powers in two randomly coupled modes traveling in opposite directions has been based on physical arguments [8, 9]. A general approach, based on the present methods, provides detailed results for various transfer function statistics of multi-layer coatings. This is possible because useful devices of the class studied here must exhibit behavior close to that of their theoretical design. We combine perturbation theory with Kronecker product methods to obtain accurate, general results for the transmission and reflection statistics of multi-layer optical coatings whose layers exhibit random optical thickness.

9.2. MATRIX ANALYSIS

For convenience set the output electric field equal to unity in Figure 9.1 and Equations (9.1) and (9.4):

$$E_t = 1. \qquad (9.6)$$

Then Equation (9.1) becomes

$$R = \frac{E_r}{E_i}, \quad T = \frac{1}{E_i}. \qquad (9.7)$$

Equations (9.2) and (9.3) are unchanged, and Equation (9.4) becomes

$$\begin{bmatrix} E_i \\ E_r \end{bmatrix} = \frac{1}{2} \begin{bmatrix} 1 & \dfrac{1}{n_i} \\ 1 & -\dfrac{1}{n_i} \end{bmatrix} \cdot \prod_{k=1}^{\ell} \mathbf{M}_k \cdot \begin{bmatrix} 1 & 1 \\ n_t & -n_t \end{bmatrix} \cdot \begin{bmatrix} 1 \\ 0 \end{bmatrix}, \quad (9.8)$$

with $\mathbf{M}_k$ given by Equation (9.5).

We determine the statistics of the transfer functions due to random deviations in layer thickness. Let

$$d_k = d_{k0} + \delta_k, \quad (9.9)$$

where d_{k0} is the design thickness of the kth layer and δ_k is its random thickness error. For the ideal design, substitute $d_k = d_{k0}$ into Equation (9.5); denote the corresponding results of Equations (9.7) and (9.2) as R_0, T_0, $\mathcal{R}_0$, and $\mathcal{T}_0$, for the ideal device.

Next, let the different layers have random thickness variations $\{\delta_k\}$ of Equation (9.9). Equations (9.8) and (9.5) become

$$\begin{bmatrix} E_i + e_i \\ E_r + e_r \end{bmatrix} = \frac{1}{2} \begin{bmatrix} 1 & \dfrac{1}{n_i} \\ 1 & -\dfrac{1}{n_i} \end{bmatrix} \cdot \prod_{k=1}^{\ell} \mathbf{M}_k \cdot \begin{bmatrix} 1 & 1 \\ n_t & -n_t \end{bmatrix} \cdot \begin{bmatrix} 1 \\ 0 \end{bmatrix}, \quad (9.10)$$

where

$$\mathbf{M}_k = \begin{bmatrix} \cos \phi_k & \dfrac{j \sin \phi_k}{n_k} \\ j n_k \sin \phi_k & \cos \phi_k \end{bmatrix}, \quad \phi_k = \frac{2\pi}{\lambda} n_k (d_{k0} + \delta_k). \quad (9.11)$$

The $\{\delta_k\}$ are assumed to be independent random variables, with specified probability densities. e_i and e_r are the departures from design values E_i and E_r, due to the random variations in layer thickness; e_i and e_r equal zero if all δ_k are zero. Then the various transfer parameters are

$$R = \frac{E_r + e_r}{E_i + e_i}, \quad T = \frac{1}{E_i + e_i}. \quad (9.12)$$

$$R = \left|\frac{E_r + e_r}{E_i + e_i}\right|^2, \quad T = \left|\frac{1}{E_i + e_i}\right|^2 \frac{n_t}{n_i}. \tag{9.13}$$

A useful device must exhibit performance close to the ideal design; the $\{\delta_k\}$ must be small enough that this is so. Consequently, Taylor series expansion of the denominators of Equations (9.12) and (9.13) is appropriate. We use these series to second order. As an example, consider the transmission transfer function T of Equation (9.12):

$$T = \frac{1}{E_i}\left[1 - \frac{e_i}{E_i} + \left(\frac{e_i}{E_i}\right)^2\right]. \tag{9.14}$$

We subsequently require expected values of T and the other quantities in Equations (9.12) and (9.13). The moments of e_i in Equation (9.14) can not be obtained directly. Therefore, we substitute

$$e_i = (E_i + e_i) - E_i \tag{9.15}$$

into Equation (9.14), to obtain

$$T = \frac{1}{E_i}\left(3 - \frac{3}{E_i}(E_i + e_i) + \frac{(E_i + e_i)^2}{E_i^2}\right). \tag{9.16}$$

The moments of $(E_i + e_i)$ are readily found by the Kronecker product methods of the following section.

The following results are similarly obtained for the other quantities of Equations (9.12) and (9.13):

$$R = \frac{(E_r + e_r)}{E_i}\left[3 - \frac{3}{E_i}(E_i + e_i) + \frac{(E_i + e_i)^2}{E_i^2}\right]. \tag{9.17}$$

$$T = \left\{\frac{6}{|E_i|^2} - \frac{8}{|E_i|^4}\text{Re}[E_i(E_i^* + e_i^*)] \right.$$
$$\left. + \frac{2}{|E_i|^6}\text{Re}[E_i^2(E_i^* + e_i^*)^2] + \frac{|E_i + e_i|^2}{|E_i|^4}\right\}\frac{n_t}{n_i}. \tag{9.18}$$

$$\mathcal{R} = |E_r + e_r|^2 \cdot \left\{ \frac{6}{|E_i|^2} - \frac{8}{|E_i|^4} \operatorname{Re}[E_i(E_i^* + e_i^*)] \right.$$

$$\left. + \frac{2}{|E_i|^6} \operatorname{Re}[E_i^2(E_i^* + e_i^*)^2] + \frac{|E_i + e_i|^2}{|E_i|^4} \right\}. \quad (9.19)$$

Finally, the two-frequency transfer function statistics yield the impulse response and signal distortion due to the random imperfections of the device, as in Sections 3.6, 3.7, 4.4, and Appendix E. Let T_1 and T_2 denote the complex transmission transfer function at free-space wavelengths λ_1 and λ_2; denote the input plane wave electric field E_i of Figure 9.1 at these two frequencies by E_{i1} and E_{i2}. Then

$$T_1 T_2^* \frac{n_t}{n_i} = \left[\frac{6}{E_{i1} E_{i2}^*} - 4\frac{E_{i2}^* + e_{i2}^*}{E_{i1} E_{i2}^{*2}} - 4\frac{E_{i1} + e_{i1}}{E_{i1}^2 E_{i2}^*} + \frac{(E_{i2}^* + e_{i2}^*)^2}{E_{i1} E_{i2}^{*3}} \right.$$

$$\left. + \frac{(E_{i1} + e_{i1})^2}{E_{i1}^3 E_{i2}^*} + \frac{(E_{i1} + e_{i1})(E_{i2}^* + e_{i2}^*)}{E_{i1}^2 E_{i2}^{*2}} \right] \frac{n_t}{n_i}. \quad (9.20)$$

For the complex reflection transfer function,

$$R_1 \cdot R_2^* = (E_{r1} + e_{r1})(E_{r2}^* + e_{r2}^*)$$

$$\cdot \left[\frac{6}{E_{i1} E_{i2}^*} - 4\frac{E_{i2}^* + e_{i2}^*}{E_{i1} E_{i2}^{*2}} - 4\frac{E_{i1} + e_{i1}}{E_{i1}^2 E_{i2}^*} + \frac{(E_{i2}^* + e_{i2}^*)^2}{E_{i1} E_{i2}^{*3}} \right.$$

$$\left. + \frac{(E_{i1} + e_{i1})^2}{E_{i1}^3 E_{i2}^*} + \frac{(E_{i1} + e_{i1})(E_{i2}^* + e_{i2}^*)}{E_{i1}^2 E_{i2}^{*2}} \right]; \quad (9.21)$$

E_{i1} and E_{i2} denote E_i at the two frequencies, as in Equation (9.20), while E_{r1} and E_{r2} denote the corresponding reflected plane wave electric field E_r of Figure 9.1.

The expected values of Equations (9.16)–(9.21) require the moments and cross-moments of the quantities $(E_i + e_i)$ and $(E_r + e_r)$ on the left-hand side of Equation (9.10). These are found by the Kronecker product methods of Section 9.3.

9.3. KRONECKER PRODUCTS

The following results, obtained from Equation (9.10), Section D.1 and Appendix C, are required to calculate the expected transmit-

9.3. KRONECKER PRODUCTS   151

tance $\langle \mathcal{T} \rangle$ from Equation (9.18), and the two-frequency transfer function statistics $\langle T_1 T_2^* \rangle$ from Equation (9.20):

$$\begin{bmatrix} \langle E_i + e_i \rangle \\ \langle E_r + e_r \rangle \end{bmatrix} = \frac{1}{2} \begin{bmatrix} 1 & \frac{1}{n_i} \\ 1 & -\frac{1}{n_i} \end{bmatrix} \cdot \prod_{k=1}^{\ell} \langle \mathbf{M}_k \rangle \cdot \begin{bmatrix} 1 & 1 \\ n_t & -n_t \end{bmatrix} \cdot \begin{bmatrix} 1 \\ 0 \end{bmatrix}. \quad (9.22)$$

$$\begin{bmatrix} \langle (E_i + e_i)^2 \rangle \\ \langle (E_i + e_i)(E_r + e_r) \rangle \\ \langle (E_r + e_r)(E_i + e_i) \rangle \\ \langle (E_r + e_r)^2 \rangle \end{bmatrix}$$

$$= \frac{1}{4} \begin{bmatrix} 1 & \frac{1}{n_i} \\ 1 & -\frac{1}{n_i} \end{bmatrix} \otimes \begin{bmatrix} 1 & \frac{1}{n_i} \\ 1 & -\frac{1}{n_i} \end{bmatrix} \cdot \prod_{k=1}^{\ell} \langle \mathbf{M}_k \otimes \mathbf{M}_k \rangle$$

$$\cdot \begin{bmatrix} 1 & 1 \\ n_t & -n_t \end{bmatrix} \otimes \begin{bmatrix} 1 & 1 \\ n_t & -n_t \end{bmatrix} \cdot \begin{bmatrix} 1 \\ 0 \\ 0 \\ 0 \end{bmatrix}. \quad (9.23)$$

$$\begin{bmatrix} \langle |E_i + e_i|^2 \rangle \\ \langle (E_i + e_i)(E_r^* + e_r^*) \rangle \\ \langle (E_r + e_r)(E_i^* + e_i^*) \rangle \\ \langle |E_r + e_r|^2 \rangle \end{bmatrix}$$

$$= \frac{1}{4} \begin{bmatrix} 1 & \frac{1}{n_i} \\ 1 & -\frac{1}{n_i} \end{bmatrix} \otimes \begin{bmatrix} 1 & \frac{1}{n_i} \\ 1 & -\frac{1}{n_i} \end{bmatrix} \cdot \prod_{k=1}^{\ell} \langle \mathbf{M}_k \otimes \mathbf{M}_k^* \rangle$$

$$\cdot \begin{bmatrix} 1 & 1 \\ n_t & -n_t \end{bmatrix} \otimes \begin{bmatrix} 1 & 1 \\ n_t & -n_t \end{bmatrix} \cdot \begin{bmatrix} 1 \\ 0 \\ 0 \\ 0 \end{bmatrix}. \quad (9.24)$$

$$\begin{bmatrix} \langle (E_{i1}+e_{i1})(E_{i2}^*+e_{i2}^*) \rangle \\ \langle (E_{i1}+e_{i1})(E_{r2}^*+e_{r2}^*) \rangle \\ \langle (E_{r1}+e_{r1})(E_{i2}^*+e_{i2}^*) \rangle \\ \langle (E_{r1}+e_{r1})(E_{r2}^*+e_{r2}^*) \rangle \end{bmatrix}$$

$$= \frac{1}{4} \begin{bmatrix} 1 & \frac{1}{n_i} \\ 1 & -\frac{1}{n_i} \end{bmatrix} \otimes \begin{bmatrix} 1 & \frac{1}{n_i} \\ 1 & -\frac{1}{n_i} \end{bmatrix} \cdot \prod_{k=1}^{\ell} \langle \mathbf{M}_k(\lambda_1) \otimes \mathbf{M}_k^*(\lambda_2) \rangle$$

$$\cdot \begin{bmatrix} 1 & 1 \\ n_t & -n_t \end{bmatrix} \otimes \begin{bmatrix} 1 & 1 \\ n_t & -n_t \end{bmatrix} \cdot \begin{bmatrix} 1 \\ 0 \\ 0 \\ 0 \end{bmatrix}. \tag{9.25}$$

The [1, 1] elements of Equations (9.22)–(9.25) appear in $\langle \mathcal{T} \rangle$ and $\langle T_1 \cdot T_2^* \rangle$. Other quantities in these Equations are not of present interest. Certain additional expected values that are required in order to calculate other transmission and reflection statistics may be found from appropriate Kronecker products; we omit these expressions here, since they are not required for the example of the following section.

The Taylor series expansions of Section 9.2 provide good approximations if

$$\frac{e_i}{E_i} \ll 1. \tag{9.26}$$

Therefore we require that

$$\sqrt{\langle |e_i|^2 \rangle} \ll |E_i|, \tag{9.27}$$

where from Equation (9.15)

$$\langle |e_i|^2 \rangle = \langle |E_i + e_i|^2 \rangle - 2 \cdot \text{Re}[E_i \langle E_i^* + e_i^* \rangle] + |E_i|^2. \tag{9.28}$$

9.4. EXAMPLE: 13-LAYER FILTER

We illustrate the present methods with a 13-layer band-pass filter centered on $\lambda = 500$ nm, with input and output media free-space.

The design parameters of Equations (9.9)–(9.11) and Figure 9.1, are [10]:

$$n_i = n_t = 1. \qquad (9.29)$$

k	1	2	3	4	5	6	7	8	9	10	11	12	13
n_k	4	1.35	4	4	1.35	4	1.35	4	1.35	4	4	1.6	4
d_{k0} (nm)[2]	31.2	92.6	31.2	31.2	92.6	31.2	92.6	31.2	92.6	31.2	31.2	78.1	31.2
ϕ_k	$\frac{\pi}{2}$	$\frac{\pi}{2}$	$\frac{\pi}{2}$	$\frac{\pi}{2}$	$\frac{\pi}{2}$	$\frac{\pi}{2}$	$\frac{\pi}{2}$	$\frac{\pi}{2}$	$\frac{\pi}{2}$	$\frac{\pi}{2}$	$\frac{\pi}{2}$	$\frac{\pi}{2}$	$\frac{\pi}{2}$

$$(9.30)$$

The design transmittance $\mathcal{T}_0$ of this filter is found from Equations (9.2), (9.5), and (9.8), and plotted in Figure 9.2.

9.4.1. Statistical Model

We assume that the random thickness parameters of Equation (9.11) are identical, uniformly distributed, independent random variables, with probability density

$$p(\delta) = \begin{cases} \dfrac{1}{2\delta_{\max}}, & |\delta| < \delta_{\max}. \\ 0, & |\delta| > \delta_{\max}. \end{cases} \qquad (9.31)$$

While the different layers have design thickness varying by almost 3 to 1, they have the same maximum thickness error. This model would correspond to a manufacturing process in which the measurement error is independent of the thickness of the layer under study. These assumptions are made only for convenience; other models are readily accommodated in the following analysis.

We require the matrix expected values on the right-hand sides of Equations (9.22)–(9.25), for $\mathbf{M}_k$ of Equation (9.11) and $p(\delta)$ of Equation (9.31). These are used in the following calculations for the expected transmittance and for the two-frequency transmission

---

[2] These values are approximate. The exact values, given by $d_k = 125/n_k$ corresponding to $\phi_k = \pi/2$ for this design, are used throughout the following calculations.

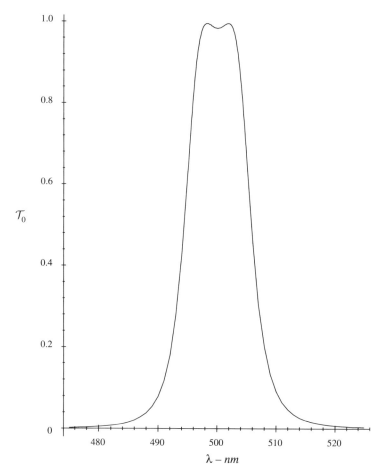

FIGURE 9.2. Design transmittance for 13-layer band-pass filter.

statistics. From Equations (9.11), (9.30), and (9.31),

$$\langle \mathbf{M}_k \rangle = \begin{bmatrix} 0 & j\dfrac{\lambda}{2\pi n_k^2 \delta_{\max}} \sin \dfrac{2\pi n_k \delta_{\max}}{\lambda} \\ j\dfrac{\lambda}{2\pi \delta_{\max}} \sin \dfrac{2\pi n_k \delta_{\max}}{\lambda} & 0 \end{bmatrix}. \quad (9.32)$$

We omit similar results for $\langle \mathbf{M}_k \otimes \mathbf{M}_k \rangle$, $\langle \mathbf{M}_k \otimes \mathbf{M}_k^* \rangle$, and $\langle \mathbf{M}_k(\lambda_1) \otimes \mathbf{M}_k^*(\lambda_2) \rangle$. All these quantities are calculated by MAPLE for the following results.

From Equation (9.28),

$$\left\langle \left| \frac{e_i}{E_i} \right|^2 \right\rangle = 1 - 2\frac{\text{Re}[E_i \langle E_i^* + e_i^* \rangle]}{|E_i|^2} + \frac{\langle |E_i + e_i|^2 \rangle}{|E_i|^2}. \qquad (9.33)$$

The expected values in the second and third terms of Equation (9.33) are given by the [1, 1] elements of Equations (9.22) and (9.24).

We choose

$$\delta_{\max} = 0.1 \text{ nm} \qquad (9.34)$$

for the examples of the following two sections; i.e., the different layers of Equation (9.30) have independent, uniformly distributed, thickness variations with maximum deviation ± 0.1 nm. For this case $\sqrt{\langle |e_i/E_i|^2 \rangle}$ varies from a minimum of approximately 0.025 at the band edges to a maximum of approximately 0.1 near the band center. The second-order Taylor series of Equations (9.14)–(9.21) should provide reasonable approximations in this case.

9.4.2. Transmittance

Random thickness errors cause the average transmittance $\langle T \rangle$ to depart from the design value T_0 of Figure 9.2. Figure 9.3 shows this deviation $T_0 - \langle T \rangle$, determined from the expected value of the second-order series expansion of Equation (9.18) and the results of Section 9.3, with random thickness errors having parameters of Equations (9.31) and (9.34).

The in-band transmittance has been reduced by a small amount near the band center, and increased slightly at the band edges. Other results, not given here, show that the ratio $\langle T \rangle / T_0$ is very close to 1 over the entire band 475–525 nm. Therefore the tolerances have not reduced the out-of-band discrimination significantly. Calculation of the in-band signal distortion requires in addition the two-frequency transmission statistics, discussed in Section 9.4.3.

9.4.3. Two-Frequency Transmission Statistics

The average transmission transfer function is obtained from Equation (9.16) as

$$\langle T \rangle = \frac{1}{E_i}\left[3 - \frac{3}{E_i}\langle E_i + e_i \rangle + \frac{\langle (E_i + e_i)^2 \rangle}{E_i^2} \right], \qquad (9.35)$$

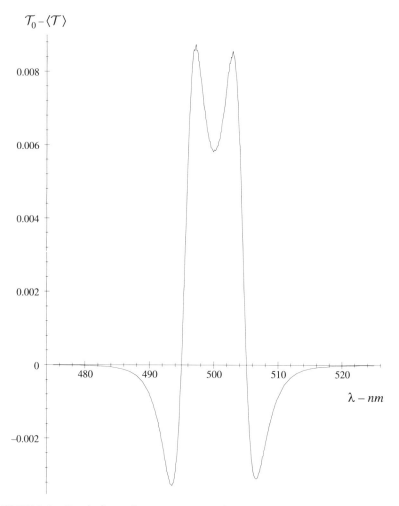

FIGURE 9.3. Deviation of average transmittance from design value. $\delta_{max} = 0.1$ nm.

using the results of Section 9.3. The a.c. component of T is

$$\Delta T = T - \langle T \rangle. \tag{9.36}$$

The variance of the transfer function fluctuation,

$$\langle |\Delta T|^2 \rangle = \langle |T|^2 \rangle - |\langle T \rangle|^2 = \langle \mathcal{T} \rangle - |\langle T \rangle|^2, \tag{9.37}$$

is plotted in Figure 9.4 for the mid-band region of the filter.

9.4. EXAMPLE: 13-LAYER FILTER

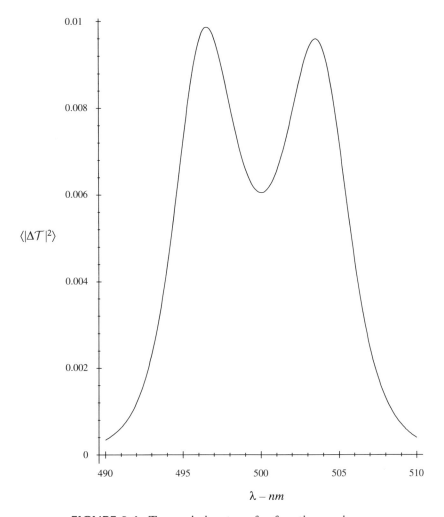

FIGURE 9.4. Transmission transfer function variance.

Finally, the covariance is given by

$$\langle \Delta T_1 \cdot \Delta T_2^* \rangle = \langle T_1 T_2^* \rangle - \langle T_1 \rangle \langle T_2 \rangle^*, \tag{9.38}$$

with $\langle T_1 T_2^* \rangle$ given by the expected value of Equation (9.20). The two-frequency correlation coefficient is given by

$$C_{12} = \frac{\langle \Delta T_1 \cdot \Delta T_2^* \rangle}{\sqrt{\langle |\Delta T_1|^2 \rangle} \cdot \sqrt{\langle |\Delta T_2|^2 \rangle}}. \tag{9.39}$$

Consider the special case

$$\lambda_1 = 500 + \Delta, \qquad \lambda_2 = 500 - \Delta, \qquad 0 < \Delta < 10 \text{ nm}, \qquad (9.40)$$

i.e., the correlation coefficient centered on 500 nm. C_{12} is so close to 1 over the mid-band region that a plot is uninteresting; it varies from its maximum value of 1 to a minimum of 0.986. This suggests that the transfer function fluctuations for different λ are strongly correlated.

9.5. DISCUSSION

Direct application of Kronecker product methods to multi-layer coatings would yield only statistics of the reciprocal transmission transfer function. Combining these methods with perturbation theory yields transmission and reflection transfer-function statistics for small departures from the ideal design. This requires that the square root of the parameter of Equation (9.33) be small compared to 1. For the 13-layer coating of Section 9.4, the parameter $\sqrt{\langle |e_i/E_i|^2 \rangle}$ has a maximum value of about 0.1 for a peak thickness variation of $\delta_{\max} = 0.1$ nm.

As $\delta_{\max}$ increases the second-order series expansions of Section 9.2 are no longer adequate. The analysis may be extended by taking higher order terms, requiring correspondingly higher order Kronecker products. However, for large enough $\delta_{\max}$ this method will fail; one hopes that this will occur beyond the region of practical interest.

For $\delta_{\max} = 1$ nm in the present 13-layer example, $\sqrt{\langle |e_i/E_i|^2 \rangle}$ is greater than 1 in the band center. The present method clearly fails in this region.

Two methods may be used to verify the present results: 1. Higher order series expansions, with higher order Kronecker products. 2. Simulation [1, 2]. If the probability density of Equation (9.31) is replaced by a discrete distribution, exact results may be obtained. For example, evaluation of 8192 matrix products of Equations (9.4) and (9.5) yields exact transfer function statistics if δ takes on only two values $\pm \delta_{\max}$ with probabilities $\frac{1}{2}$.

REFERENCES

1. Neil A. Jackman, *Limitations and Tolerances in Optical Devices*, Ph. D. Dissertation, Stevens Institute of Technology, Hoboken, NJ, 1994; Chapter 5.
2. Neil A. Jackman, "Multilayer Optical Filters with Random Errors," *Bell Labs Technical Journal*, January–March 1998, pp. 112–121.
3. Alfred Thelen, *Design of Optical Interference Coatings*, McGraw-Hill, New York, 1988.
4. Pochi Yeh, *Optical Waves in Layered Media*, Wiley, New York, 1988.
5. H. A. Macleod, *Thin-Film Optical Filters*, McGraw-Hill, New York, 1989.
6. J. A. Dobrowolski, "Numerical Methods for Optical Thin Films," *Optics and Photonics News*, June 1997, pp. 23–33.
7. Harrison E. Rowe, "Propagation in One-Dimensional Random Media," *IEEE Transactions on Microwave Theory and Techniques*, Vol. MTT-19, January 1971, pp. 73–80.
8. D. Marcuse, "Coupled Power Equations for Backward Waves," *IEEE Transactions on Microwave Theory and Techniques*, Vol. MTT-20, August 1972, pp. 541–546.
9. Dietrich Marcuse, *Theory of Dielectric Optical Waveguides*, 2nd ed., Academic Press, New York, 1991.
10. Alfred Thelen, reference [3], page 201, Figure 10.3.

CHAPTER TEN

Conclusion

The main thrust of the present work is the use of Kronecker product methods in the calculation of transmission statistics of multimode systems with random coupling. Such systems exhibit linearity between complex mode amplitudes; Kronecker products yield linear relations between corresponding moments of the mode amplitudes.

These methods require that the system must be regarded as a cascade of sections, described by a matrix product. This description is natural for systems such as multi-layer coatings. For systems with continuous coupling, a matrix description must be artificially introduced.

For forward-traveling modes with white coupling, i.e., resulting from imperfections with a very short correlation length, the system may be divided into sections of infinitesimal length; exact linear differential equations for the mode powers or for higher moments are obtained. For more general spectra, the sections must be long enough for the coupling in different sections to be approximately statistically independent, and the coupling must be small enough for perturbation theory to be applied to the individual sections; the resulting coupled power or other differential equations for the higher moments are only approximate.

Multi-layer coatings offer an important two-mode system with modes traveling in opposite directions. A combination of Kronecker product methods with perturbation theory yields approximate transfer function statistics for systems with performance close to that of the ideal design.

The use of MAPLE or an alternative program capable of symbolic computation is essential for the analysis described here.

The examples described here have been included to illustrate the possibilities of the present methods; they are not intended to be exhaustive. Other models, coupling statistics, input signal statistics may be similarly treated.

The region of validity for the approximate results that have been presented are only heuristic. Accuracy of these methods is confirmed by studying special cases that admit exact solutions. These include forward-traveling modes with random square-wave coupling having Markov statistics, and multi-layer coatings with random thickness having a discrete probability distribution. Some results for the former case have been presented here.

Finally, similar matrix methods have appeared in the physics literature [1–3].

REFERENCES

1. A. R. McGurn, K. T. Christensen, F. M. Mueller, and A. A. Maradudin, "Anderson localization in one-dimensional randomly disordered optical systems that are periodic on average," *Physical Review B*, Vol. 47, 15 May 1993-II, pp. 13 120–13 125.
2. J. B. Pendry, "Symmetry and transport of waves in one-dimensional disordered systems," *Advances in Physics*, Vol. 43, 1994, pp. 461–542.
3. V. D. Freilikher, B. A. Liansky, I. V. Yurkevich, A. A. Maradudin, and A. R. McGurn, "Enhanced transmission due to disorder," *Physical Review E*, Vol. 51, June 1995, pp. 6301–6304.

APPENDIX A

Series Solution for the Coupled Line Equations

We apply the method of successive approximations (Picard's method) [1, 2] to the normalized coupled line equations for two modes, given in Equations (2.32)–(2.34) [3]. Recall that $I_0(z)$ and $G_0(z)$ denote the signal mode, $I_1(z)$ and $G_1(z)$ denote the spurious mode. The signal mode has lower loss; consequently,

$$\Delta\alpha(z) = \alpha_0(z) - \alpha_1(z) \leq 0, \qquad \operatorname{Re}\Delta\gamma(z) = \int_0^z \Delta\alpha(x)\,dx < 0. \quad (A.1)$$

Denote the exact solution to Equation (2.33) by

$$\begin{bmatrix} G_0(z) \\ G_1(z) \end{bmatrix} = \mathbf{P} \cdot \begin{bmatrix} G_0(0) \\ G_1(0) \end{bmatrix} = \begin{bmatrix} p_{00} & p_{01} \\ p_{10} & p_{11} \end{bmatrix} \cdot \begin{bmatrix} G_0(0) \\ G_1(0) \end{bmatrix}. \quad (A.2)$$

The method of successive approximations yields infinite series for the elements of $\mathbf{P}$, of which the first two terms are the perturbation solution $\mathbf{M}$ of Equations (2.35) and (2.36).

Consider first the initial conditions

$$G_0(0) = 1, \qquad G_1(0) = 0, \quad (A.3)$$

in order to obtain p_{00} and p_{10}. Let $G_{0(n)}(z)$ and $G_{1(n)}(z)$ be the nth approximation to the solution of Equation (2.33). Let the initial approximation be given simply by the initial conditions of Equation

(A.3). The successive approximations are:

$$G_{0(0)}(z) = 1. \quad G_{1(0)}(z) = 0.$$

$$G_{0(1)}(z) = 1.$$

$$G_{1(1)}(z) = j \int_0^z c(x) e^{-\Delta\gamma(x)} dx.$$

$$G_{0(2)}(z) = 1 - \int_0^z c(x) e^{+\Delta\gamma(x)} dx \int_0^x c(y) e^{-\Delta\gamma(y)} dy.$$

$$G_{1(2)} = j \int_0^z c(x) e^{-\Delta\gamma(x)} dx.$$

(A.4)

$$\cdots\cdots\cdots$$

$$G_{0(n)}(z) = 1 + j \int_0^z c(x) e^{+\Delta\gamma(x)} G_{1(n-1)}(x) dx.$$

$$G_{1(n)}(z) = j \int_0^z c(x) e^{-\Delta\gamma(x)} G_{0(n-1)}(x) dx.$$

The nth approximation is obtained by substituting the $(n-1)$th approximation in the right-hand side of Equation (2.33) and integrating. Define

$$G_{0(n)}(z) - G_{0(n-1)}(z) = p_{00(n)}(z),$$
$$G_{1(n)}(z) - G_{1(n-1)}(z) = p_{10(n)}(z).$$

(A.5)

Then,

$$G_{0(n)}(z) = 1 + \sum_{k=1}^n p_{00(k)}(z),$$

$$G_{1(n)}(z) = \sum_{k=1}^n p_{10(k)}(z).$$

(A.6)

The terms of Equation (A.6) satisfy the following relations:

$$p_{00(n)}(z) = j \int_0^z c(x) e^{+\Delta\gamma(x)} p_{10(n-1)}(x) dx, \quad n \geq 1.$$

$$p_{10(n)}(z) = j \int_0^z c(x) e^{-\Delta\gamma(x)} p_{00(n-1)}(x) dx, \quad n \geq 1.$$

(A.7)

$$p_{00(0)}(z) = 1. \quad p_{10(0)}(z) = 0.$$

The $\lim_{n\to\infty}$ of the summations in Equation (A.6) yields the first column of the matrix **P** in Equation (A.2):

$$p_{00} = 1 + \sum_{n=1}^{\infty} p_{00(n)}(z).$$

$$p_{10} = \sum_{n=1}^{\infty} p_{10(n)}(z). \tag{A.8}$$

Note that

$$p_{00(n)}(z) = 0, \quad n \text{ odd}; \quad p_{10(n)}(z) = 0, \quad n \text{ even}. \tag{A.9}$$

The first and second summations of Equations (A.6) and (A.8) contain only even and odd terms, respectively.

We obtain the following bounds on the terms of Equations (A.6) and (A.8):

$$|p_{00(n)}(z)| \begin{cases} \leq \dfrac{[\int_0^z |c(x)|dx]^n}{n!}, & n \text{ even}. \\ = 0, & n \text{ odd}. \end{cases} \tag{A.10}$$

$$|p_{10(n)}(z)| \begin{cases} = 0, & n \text{ even}. \\ \leq \dfrac{[\int_0^z |c(x)|dx]^n}{n!} e^{-\int_0^z \Delta\alpha(x)dx}, & n \text{ odd}. \end{cases} \tag{A.11}$$

Let Equation (A.10) hold for some even n. Then from the middle relation of Equation (A.7),

$$|p_{10(n+1)}(z)| \leq \int_0^z |c(x)| e^{-\int_0^x \Delta\alpha(y)dy} \frac{[\int_0^x |c(y)|dy]^n}{n!} dx$$

$$\leq \frac{e^{-\int_0^z \Delta\alpha(y)dy}}{n!} \int_0^z \left[\int_0^x |c(y)|dy\right]^n d\left[\int_0^x |c(y)|dy\right]$$

$$= \frac{[\int_0^z |c(x)|dx]^{n+1}}{(n+1)!} e^{-\int_0^z \Delta\alpha(x)dx}. \tag{A.12}$$

From Equation (A.12) and the first relation of Equation (A.7),

$$|p_{00(n+2)}(z)| \le \int_0^z |c(x)| e^{+\int_0^x \Delta\alpha(y)dy} \frac{[\int_0^x |c(y)|dy]^{n+1}}{(n+1)!} e^{-\int_0^x \Delta\alpha(y)dy} dx$$

$$= \frac{1}{(n+1)!} \int_0^z \left[\int_0^x |c(y)|dy\right]^{n+1} d\left[\int_0^x |c(y)|dy\right]$$

$$= \frac{[\int_0^z |c(x)|dx]^{n+2}}{(n+2)!}. \qquad (A.13)$$

The bounds of Equations (A.10) and (A.11) hold for all n by induction.

The exact solutions for two degenerate modes, Equation (2.8), and for a discrete coupler, Equation (2.11), attain the limits given in Equations (A.10) and (A.11). Therefore these bounds cannot be improved without additional restrictions on the coupling coefficient $c(z)$.

Similar results for the second column of the matrix **P** in Equation (A.2) are readily obtained:

$$p_{01} = \sum_{n=1}^{\infty} p_{01(n)}(z).$$

$$p_{11} = 1 + \sum_{n=1}^{\infty} p_{11(n)}(z). \qquad (A.14)$$

$$|p_{11(n)}(z)| \begin{cases} \le \dfrac{[\int_0^z |c(x)|dx]^n}{n!} e^{-\int_0^z \Delta\alpha(x)dx}, & n \text{ even}, n \ge 2. \\ = 0, & n \text{ odd}. \end{cases} \qquad (A.15)$$

$$|p_{01(n)}(z)| \begin{cases} = 0, & n \text{ even}. \\ \le \dfrac{[\int_0^z |c(x)|dx]^n}{n!}, & n \text{ odd}. \end{cases} \qquad (A.16)$$

$$p_{11(0)}(z) = 1; \qquad p_{01(0)}(z) = 0. \qquad (A.17)$$

The multi-mode case can be treated in a similar manner. We introduce compact notation for Equation (2.69):

$$\mathbf{G}'(z) = jc(z)\mathbf{K}(z) \cdot \mathbf{G}(z), \qquad (A.18)$$

SERIES SOLUTION FOR THE COUPLED LINE EQUATIONS

where

$$\mathbf{K}(z) = e^{+\gamma(z)} \cdot \mathbf{C} \cdot e^{-\gamma(z)}. \tag{A.19}$$

Define

$$\Delta\gamma_{ij}(z) = \int_0^z \{\Gamma_i(z) - \Gamma_j(z)\}dz. \tag{A.20}$$

Then writing out the matrix $\mathbf{K}(z)$,

$$\mathbf{K}(z) = \begin{bmatrix} 0 & e^{\Delta\gamma_{01}(z)}C_{01} & e^{\Delta\gamma_{02}(z)}C_{02} & \cdots \\ e^{-\Delta\gamma_{01}(z)}C_{01} & 0 & e^{\Delta\gamma_{12}(z)}C_{12} & \cdots \\ e^{-\Delta\gamma_{02}(z)}C_{02} & e^{-\Delta\gamma_{12}(z)}C_{12} & 0 & \cdots \\ \vdots & \vdots & \vdots & \ddots \end{bmatrix}. \tag{A.21}$$

The method of successive approximations yields

$$\mathbf{G}(z) = \mathbf{G}(0) + \sum_{n=1}^{\infty} \mathbf{P}_{(n)}(z), \tag{A.22}$$

where

$$\mathbf{P}_{(n)}(z) = j\int_0^z c(x)\mathbf{K}(x) \cdot \mathbf{P}_{(n-1)}(x)dx, \qquad \mathbf{P}_{(0)}(z) = \mathbf{G}(0). \tag{A.23}$$

Truncating the summation of Equation (A.22) at $n = 2$ yields the perturbation solution of Equation (2.70). Bounds on the terms of the summation in Equation (A.22) for the lossless case are as follows:

$$\|\mathbf{P}_{(n)}(z)\| < \frac{[\int_0^z |c(x)|dx]^n}{n!}\|\mathbf{C}\|^n \|\mathbf{G}(0)\|,$$
$$\alpha_i(z) = 0, \quad i = 0, 1, 2, \cdots, \tag{A.24}$$

where $\|\ \|$ denotes the vector or matrix norm [4, 5];

$$\|\mathbf{C}\| = \max_j \sum_i |C_{ij}|; \qquad \|\mathbf{G}(0)\| = \sum_i |G_i(0)| = \sum_i |I_i(0)|. \tag{A.25}$$

REFERENCES

1. E. L. Ince, *Ordinary Differential Equations*, Dover Publications, New York, 1956.
2. R. Bellman, *Stability Theory of Differential Equations*, McGraw-Hill, New York, 1953.
3. H. E. Rowe, "Approximate Solutions for the Coupled Line Equations," *Bell System Technical Journal*, Vol. 41, May 1962, pp. 1011–1029.
4. Assem S. Deif, *Advanced Matrix Theory for Scientists and Engineers*, Wiley, New York, 1982.
5. Gilbert Strang, *Linear Algebra and its Applications*, 2nd ed., Academic Press, New York, 1980.

APPENDIX B

General Transmission Properties of Two-Mode Guide

We obtain general transmission properties for the signal mode transfer function and impulse response of two-mode guide from the exact solution given in Appendix A [1]. The infinite series solution for the G_0, given in Equations (A.6)–(A.9), takes on the following form for a guide of length L with constant propagation functions:

$$G_0(\Delta\Gamma) = 1 + \sum_{n=1}^{\infty} p_{00(2n)}(\Delta\Gamma), \qquad (B.1)$$

where $p_{00(n)}(\ )$ is given in Equation (A.7), and may be rewritten as

$$p_{00(2n)}(\Delta\Gamma) = (-1)^n \int_0^L dx_1 \int_0^{x_1} dx_2 \cdots \int_0^{x_{2n-1}} dx_{2n}$$
$$\cdot c(x_1)c(x_2)\cdots c(x_{2n}) \cdot e^{\Delta\Gamma(x_1-x_2+x_3-x_4+\cdots-x_{2n})}. \qquad (B.2)$$

We have suppressed the L dependence in Equations (B.1) and (B.2), as discussed in Section 3.2. Bounds on the terms of Equation (B.1) are given in Equation (A.10):

$$|p_{00(2n)}(\Delta\Gamma)| \le \frac{[\int_0^L |c(x)|dx]^{2n}}{(2n)!}, \qquad \Delta\alpha \le 0. \qquad (B.3)$$

Under suitable conditions the individual terms $p_{00(2n)}(\Delta\Gamma)$ are analytic in the $\Delta\Gamma$ plane, the series in Equation (B.1) is uniformly convergent, and $G_0(\Delta\Gamma)$ is analytic in the $\Delta\Gamma$ plane [2].

Set

$$s = x_1 - x_2 + x_3 - x_4 + \cdots - x_{2n} \quad (B.4)$$

in Equation (B.2) to yield

$$p_{00(2n)}(\Delta\alpha, \Delta\beta) = \int_0^L f_n(s) e^{\Delta\alpha s} e^{j\Delta\beta s} ds, \quad (B.5)$$

where $f_n(s)$ is a real, $(2n-1)$-fold integral. Set

$$s = L\tau \quad (B.6)$$

in Equation (B.5), substitute into Equation (B.1), and interchange $\sum$ and $\int$; then

$$G_0(\Delta\alpha, \Delta\beta) = 1 + L\int_0^1 \left\{\sum_{n=1}^{\infty} f_n(L\tau)\right\} e^{\Delta\alpha L\tau} e^{j\Delta\beta L\tau} d\tau. \quad (B.7)$$

Finally, define

$$f(L\tau) = \sum_{n=1}^{\infty} f_n(L\tau) \quad (B.8)$$

to yield

$$G_0(\Delta\alpha, \Delta\beta) = 1 + \int_0^1 f(L\tau) e^{\Delta\alpha L\tau} e^{j\Delta\beta L\tau}. \quad (B.9)$$

Comparing Equation (B.9) with Equation (3.17),

$$g_{\Delta\alpha}(\tau) = \begin{cases} \delta(\tau) + Lf(L\tau)e^{\Delta\alpha L\tau}, & 0 \le \tau \le 1. \\ 0, & \text{otherwise.} \end{cases} \quad (B.10)$$

$g_{\Delta\alpha}(\tau)$ is causal, time-limited, and real.
Define

$$G_0(\Delta\alpha, \Delta\beta) = 1 - A(\Delta\alpha, \Delta\beta) + j\Theta(\Delta\alpha, \Delta\beta), \quad (B.11)$$

where A and Θ are real. A is the departure of $\text{Re } G_0$ from unity, and Θ is $\text{Im } G_0$, respectively. (For small coupling $A \approx$ loss and $\Theta \approx$ phase, but we do not make these approximations here.) Then

Equation (B.10) yields the following four properties of $G_0(\Delta\alpha, \Delta\beta)$ and $g_{\Delta\alpha}(\tau)$:

1. A and Θ are Hilbert transforms:

$$\Theta = \widehat{A}, \qquad (B.12)$$

where $\widehat{}$ denotes the Hilbert transform,

$$\widehat{A}(\Delta\alpha, \Delta\beta) = \frac{1}{\pi} A(\Delta\alpha, \Delta\beta) \star \frac{1}{\Delta\beta}$$

$$= \frac{1}{\pi} \int_{-\infty}^{\infty} \frac{A(\Delta\alpha, \sigma)}{\Delta\beta - \sigma} d\sigma, \qquad (B.13)$$

and $\star$ represents the convolution operator. The inverse relationship is

$$A = \widehat{\Theta} + \overline{A}, \qquad \overline{A} = \lim_{B \to \infty} \frac{1}{2B} \int_{-B}^{B} A(\Delta\alpha, \sigma) d\sigma. \qquad (B.14)$$

$\overline{A}$ represents the d.c. component of A.

2. From Equation (B.10),

$$g_{\Delta\alpha}(\tau) = e^{-|\Delta\alpha|L\tau} g_{\Delta\alpha=0}(\tau), \qquad \Delta\alpha \le 0. \qquad (B.15)$$

Therefore G_0 and $g_{\Delta\alpha}$ for any negative $\Delta\alpha$ are related to the corresponding quantities for $\Delta\alpha = 0$ as follows:

$$G_0(\Delta\alpha, \Delta\beta) = \frac{1}{2\pi} G_0(0, \Delta\beta) \star \frac{1}{|\Delta\alpha| - j\Delta\beta}. \qquad (B.16)$$

$$A(\Delta\alpha, \Delta\beta) = \frac{1}{\pi} A(0, \Delta\beta) \star \frac{|\Delta\alpha|}{\Delta\alpha^2 + \Delta\beta^2}. \qquad (B.17)$$

$$\Theta(\Delta\alpha, \Delta\beta) = \frac{1}{\pi} \Theta(0, \Delta\beta) \star \frac{|\Delta\alpha|}{\Delta\alpha^2 + \Delta\beta^2}. \qquad (B.18)$$

Equations (B.17) and (B.18) show that as the differential loss increases ($\Delta\alpha$ becomes more negative), the transfer function fluctuations are reduced. The window function of these relations has unit area:

$$\frac{1}{\pi|\Delta\alpha|} \int_{-\infty}^{\infty} \frac{1}{1 + \left(\frac{\Delta\beta}{\Delta\alpha}\right)^2} d(\Delta\beta) = 1. \quad \text{(B.19)}$$

This window function must approach a unit impulse for small $\Delta\alpha$:

$$\lim_{\Delta\alpha \to 0} \frac{1}{\pi|\Delta\alpha|} \frac{1}{1 + \left(\frac{\Delta\beta}{\Delta\alpha}\right)^2} = \delta(\Delta\beta). \quad \text{(B.20)}$$

3. G_0 is strictly band-limited, with sample points

$$\Delta\beta_n = \frac{2\pi n}{L}. \quad \text{(B.21)}$$

Then,

$$G_0(\Delta\alpha, \Delta\beta) = e^{j\pi\left(\frac{\Delta\beta L}{2\pi}\right)} \sum_{n=-\infty}^{\infty} G_0\left(\Delta\alpha, \frac{2\pi n}{L}\right)(-1)^n \frac{\sin \pi\left(\frac{\Delta\beta L}{2\pi} - n\right)}{\pi\left(\frac{\Delta\beta L}{2\pi} - n\right)}. \quad \text{(B.22)}$$

4. G_0 is Hermitian, A symmetric, and Θ anti-symmetric:

$$G_0(\Delta\alpha, -\Delta\beta) = G_0^*(\Delta\alpha, \Delta\beta).$$
$$A(\Delta\alpha, -\Delta\beta) = A(\Delta\alpha, \Delta\beta). \quad \text{(B.23)}$$
$$\Theta(\Delta\alpha, -\Delta\beta) = -\Theta(\Delta\alpha, \Delta\beta).$$

$G_0(\Delta\Gamma)$ is analytic in the entire $\Delta\Gamma$ plane, and therefore has no poles. Results for the location of 0's of $G_0(\Delta\Gamma)$ yield sufficient conditions on $\int_{-\infty}^{\infty} |c(x)| dx$ that guarantee G_0 will be minimum-phase [3]. We omit these results, since they do not appear in the present work.

All of the present results are exact; in particular they are not restricted to the small-coupling regime required by perturbation theory.

REFERENCES

1. Harrison E. Rowe and D. T. Young, "Transmission Distortion in Multimode Random Waveguides," *IEEE Transactions on Microwave Theory and Techniques*, Vol. MTT-20, June 1972, pp. 349–365.
2. E. A. Coddington and N. Levinson, *Theory of Ordinary Differential Equations*, McGraw-Hill, New York, 1955; p. 40, prob. 7.
3. Harrison E. Rowe and D. T. Young, "Minimum-Phase Behavior of Random Media," *IEEE Transactions on Microwave Theory and Techniques*, Vol. MTT-23, May 1975, pp. 411–416.

APPENDIX C

Kronecker Products

Consider two arbitrary matrices

$$\mathbf{A} = \begin{bmatrix} a_{11} & a_{12} & \cdots \\ a_{21} & a_{22} & \cdots \\ \vdots & \vdots & \ddots \end{bmatrix}, \qquad \mathbf{B} = \begin{bmatrix} b_{11} & b_{12} & \cdots \\ b_{21} & b_{22} & \cdots \\ \vdots & \vdots & \ddots \end{bmatrix}, \qquad (C.1)$$

not necessarily square. The ordinary matrix product of two conformable matrices is denoted by $\mathbf{A} \cdot \mathbf{B}$; the numbers of columns of $\mathbf{A}$ and rows of $\mathbf{B}$ must match. The Kronecker product of these two matrices is denoted by [1, 2]

$$\mathbf{A} \otimes \mathbf{B} = \begin{bmatrix} a_{11}\mathbf{B} & a_{12}\mathbf{B} & \cdots \\ a_{21}\mathbf{B} & a_{22}\mathbf{B} & \cdots \\ \vdots & \vdots & \ddots \end{bmatrix}. \qquad (C.2)$$

There is no restriction on the dimensions of the two matrices. A representative submatrix of Equation (C.2) is

$$a_{ij}\mathbf{B} = \begin{bmatrix} a_{ij}b_{11} & a_{ij}b_{12} & \cdots \\ a_{ij}b_{21} & a_{ij}b_{22} & \cdots \\ \vdots & \vdots & \ddots \end{bmatrix}. \qquad (C.3)$$

The Kronecker product $\mathbf{A} \otimes \mathbf{B}$ has dimensions equal to the products of the dimensions of $\mathbf{A}$ and $\mathbf{B}$.

The following properties are used throughout the present text [1, 2]:

1. Both Kronecker products and ordinary matrix products are not commutative:

$$\mathbf{A} \otimes \mathbf{B} \neq \mathbf{B} \otimes \mathbf{A}. \tag{C.4}$$

2. Kronecker products are associative and distributive:

$$\mathbf{A} \otimes \mathbf{B} \otimes \mathbf{C} = (\mathbf{A} \otimes \mathbf{B}) \otimes \mathbf{C} = \mathbf{A} \otimes (\mathbf{B} \otimes \mathbf{C}). \tag{C.5}$$

$$(\mathbf{A}+\mathbf{B}) \otimes (\mathbf{C}+\mathbf{D}) = \mathbf{A} \otimes \mathbf{C} + \mathbf{A} \otimes \mathbf{D} + \mathbf{B} \otimes \mathbf{C} + \mathbf{B} \otimes \mathbf{D}. \tag{C.6}$$

Both $\mathbf{A}$ and $\mathbf{B}$ must be the same size, and $\mathbf{C}$ and $\mathbf{D}$ must be the same (possibly different) size in Equation (C.6).

3. Interchange of Kronecker and ordinary matrix products is essential to the present work:

$$(\mathbf{A} \otimes \mathbf{B}) \cdot (\mathbf{C} \otimes \mathbf{D}) = (\mathbf{A} \cdot \mathbf{C}) \otimes (\mathbf{B} \cdot \mathbf{D}). \tag{C.7}$$

Both $\mathbf{A}$ and $\mathbf{C}$ must be conformable, and $\mathbf{B}$ and $\mathbf{D}$ must be conformable, in Equation (C.7). This relation is readily generalized;

$$(\mathbf{A}_1 \otimes \mathbf{A}_2 \otimes \mathbf{A}_3 \otimes \cdots) \cdot (\mathbf{B}_1 \otimes \mathbf{B}_2 \otimes \mathbf{B}_3 \otimes \cdots)$$
$$\cdot (\mathbf{C}_1 \otimes \mathbf{C}_2 \otimes \mathbf{C}_3 \otimes \cdots) \cdot \cdots$$
$$= (\mathbf{A}_1 \cdot \mathbf{B}_1 \cdot \mathbf{C}_1 \cdot \cdots) \otimes (\mathbf{A}_2 \cdot \mathbf{B}_2 \cdot \mathbf{C}_2 \cdot \cdots)$$
$$\otimes (\mathbf{A}_3 \cdot \mathbf{B}_3 \cdot \mathbf{C}_3 \cdot \cdots) \otimes \cdots. \tag{C.8}$$

The matrix dimensions in each of the parentheses in the second line of Equation (C.8) must satisfy the conformability requirements.

4. Inverse and transpose:

$$(\mathbf{A} \otimes \mathbf{B})^{-1} = \mathbf{A}^{-1} \otimes \mathbf{B}^{-1}. \tag{C.9}$$

$$(\mathbf{A} \otimes \mathbf{B})^{\mathrm{T}} = \mathbf{A}^{\mathrm{T}} \otimes \mathbf{B}^{\mathrm{T}}. \tag{C.10}$$

Both $\mathbf{A}$ and $\mathbf{B}$ must be square in Equation (C.9). These relations generalize in an obvious way.

5. Denote the eigenvalues and eigenvectors of **A** and **B** by α_i, β_j and $\mathbf{a}_i, \mathbf{b}_j$, respectively,

$$\mathbf{A} \cdot \mathbf{a}_i = \alpha_i \mathbf{a}_i, \qquad \mathbf{B} \cdot \mathbf{b}_j = \beta_j \mathbf{b}_j. \qquad (C.11)$$

Define γ_k and $\mathbf{c}_k$ by

$$(\mathbf{A} \otimes \mathbf{B}) \cdot \mathbf{c}_k = \gamma_k \mathbf{c}_k. \qquad (C.12)$$

Then from Equation (C.7)

$$\gamma_k = \alpha_i \beta_j, \qquad \mathbf{c}_k = \mathbf{a}_i \otimes \mathbf{b}_j. \qquad (C.13)$$

The eigenvalues of $\mathbf{A} \otimes \mathbf{B}$ are products of all pairs of eigenvalues, and the eigenvectors of $\mathbf{A} \otimes \mathbf{B}$ are Kronecker products of all pairs of eigenvectors, of **A** and **B**.

6. Let **a** be a column vector and $\mathbf{b}^T$ be a row vector.

$$\mathbf{a} \cdot \mathbf{b}^T = \mathbf{a} \otimes \mathbf{b}^T = \mathbf{b}^T \otimes \mathbf{a}. \qquad (C.14)$$

REFERENCES

1. R. Bellman, *Introduction to Matrix Analysis*, McGraw-Hill, New York, 1970; Chapters 12 and 15.
2. F. A. Graybill, *Introduction to Matrices with Applications to Statistics*, Wadsworth, Belmont, California, 1969; Section 8.8

APPENDIX D

Expected Values of Matrix Products

D.1. INDEPENDENT MATRICES

Consider the scalar problem

$$I(n) = M(n)M(n-1)\cdots M(2)M(1)I(0), \qquad (D.1)$$

where the $M(k)$ are independent random variables. For example, this relation may represent a chain of n noninteracting amplifiers with random independent gains, with input $I(0)$ and output $I(n)$. The expected output is

$$\langle I(n)\rangle = \langle M(n)\rangle\langle M(n-1)\rangle \cdots \langle M(2)\rangle\langle M(1)\rangle\langle I(0)\rangle. \qquad (D.2)$$

Assume the different $M(k)$ have the same statistics; then

$$\langle I(n)\rangle = \langle M\rangle^n \langle I(0)\rangle. \qquad (D.3)$$

Similarly the expected output power is

$$\langle |I(n)|^2\rangle = \langle |M|^2\rangle^n \langle |I(0)|^2\rangle. \qquad (D.4)$$

We require the matrix extension of these elementary results for the N-mode case. First, the expected value of a matrix is defined

180 EXPECTED VALUES OF MATRIX PRODUCTS

as the matrix of expected values of its elements; i.e., the expected value of

$$\mathbf{A} = \begin{bmatrix} a_{11} & a_{12} & \cdots \\ a_{21} & a_{22} & \cdots \\ \vdots & \vdots & \ddots \end{bmatrix} \quad (D.5)$$

is

$$\langle \mathbf{A} \rangle = \begin{bmatrix} \langle a_{11} \rangle & \langle a_{12} \rangle & \cdots \\ \langle a_{21} \rangle & \langle a_{22} \rangle & \cdots \\ \vdots & \vdots & \ddots \end{bmatrix}. \quad (D.6)$$

Consider an N-mode guide divided into n sections. Denote the column vector of mode complex amplitudes after k sections by $I(k)$:

$$I^T(k) = [I_0(k) \quad I_1(k) \quad I_2(k) \quad \cdots \quad I_{N-1}(k)]. \quad (D.7)$$

Let $\mathbf{M}(k)$ denote the wave matrix describing the kth section:

$$I(k) = \mathbf{M}(k) \cdot I(k-1). \quad (D.8)$$

Then the vector output after n sections is

$$I(n) = \mathbf{M}(n) \cdot \mathbf{M}(n-1) \cdots \mathbf{M}(2) \cdot \mathbf{M}(1) \cdot I(0)$$
$$= \prod_{k=1}^{n} \mathbf{M}(n+1-k) \cdot I(0), \quad (D.9)$$

where $I(0)$ is the vector input, and the $\mathbf{M}(k)$ are $N \times N$ random matrices.

Assume different $\mathbf{M}(k)$ are independent. Any element of $\mathbf{M}(k)$ is statistically independent of any element of $\mathbf{M}(r)$, $k \neq r$; however, different elements of any particular matrix are not independent, e.g., see Equation (3.24). Denote the elements of $\mathbf{M}(k)$ by $m_{ij}(k)$; then $m_{ij}(k)$ is independent of $m_{pq}(r)$ if $k \neq r$ for any i, j, p, q. The ℓth element of Equation (D.9) is

$$I_\ell(n) = \sum_{k_1 \cdots k_n} m_{\ell k_1}(n) m_{k_1 k_2}(n-1) \cdots m_{k_{n-1} k_n} I_{k_n}(0). \quad (D.10)$$

D.1. INDEPENDENT MATRICES

Since different m's in each term of the summation of Equation (D.10) are statistically independent, the expected value of Equation (D.10) yields the average outputs:

$$\langle I_\ell(n) \rangle = \sum_{k_1 \cdots k_n} \langle m_{\ell k_1}(n) \rangle \langle m_{k_1 k_2}(n-1) \rangle \cdots \langle m_{k_{n-1} k_n} \rangle \langle I_{k_n}(0) \rangle. \quad (D.11)$$

In matrix notation, Equation (D.11) becomes

$$\langle I(n) \rangle = \langle \mathbf{M}(n) \rangle \cdot \langle \mathbf{M}(n-1) \rangle \cdot \cdots \cdot \langle \mathbf{M}(2) \rangle \cdot \langle \mathbf{M}(1) \rangle \cdot \langle I(0) \rangle$$
$$= \prod_{k=1}^{n} \langle \mathbf{M}(n+1-k) \rangle \cdot \langle I(0) \rangle. \quad (D.12)$$

The alternative result

$$\langle I(k) \rangle = \langle \mathbf{M}(k) \rangle \cdot \langle I(k-1) \rangle \quad (D.13)$$

follows from Equation (D.12). Finally, if the $\mathbf{M}(k)$ have the same statistics,

$$\langle I(n) \rangle = \langle \mathbf{M} \rangle^n \cdot \langle I(0) \rangle,$$
$$\langle I(k) \rangle = \langle \mathbf{M} \rangle \cdot \langle I(k-1) \rangle. \quad (D.14)$$

Second- and higher order statistics are obtained by the use of Kronecker products, described in Appendix C [1–3]. For second order, take the Kronecker product of Equation (D.9) with its complex conjugate, and use Equation (C.8) to obtain

$$I(n) \otimes I^*(n) = \prod_{k=1}^{n} [\mathbf{M}(n+1-k) \otimes \mathbf{M}^*(n+1-k)] \cdot [I(0) \otimes I^*(0)],$$
$$(D.15)$$

where the $\prod$ denotes ordinary ($\cdot$) matrix multiplication; the $\mathbf{M}(k) \otimes \mathbf{M}^*(k)$ are $N^2 \times N^2$ random matrices. The expected value of Equation (D.15) is obtained by the same argument used above for the

182 EXPECTED VALUES OF MATRIX PRODUCTS

average output:

$$\langle I(n) \otimes I^*(n) \rangle = \prod_{k=1}^{n} \langle \mathbf{M}(n+1-k) \otimes \mathbf{M}^*(n+1-k) \rangle \cdot \langle I(0) \otimes I^*(0) \rangle. \tag{D.16}$$

If the $\mathbf{M}(k)$ have the same statistics,

$$\langle I(n) \otimes I^*(n) \rangle = \langle \mathbf{M} \otimes \mathbf{M}^* \rangle^n \cdot \langle I(0) \otimes I^*(0) \rangle,$$
$$\langle I(k) \otimes I^*(k) \rangle = \langle \mathbf{M} \otimes \mathbf{M}^* \rangle \cdot \langle I(k-1) \otimes I^*(k-1) \rangle. \tag{D.17}$$

We write out Equation (D.17) explicitly for the two-mode case:

$$\begin{bmatrix} \langle |I_0(n)|^2 \rangle \\ \langle I_0(n)I_1^*(n) \rangle \\ \langle I_1(n)I_0^*(n) \rangle \\ \langle |I_1(n)|^2 \rangle \end{bmatrix} = \langle \mathbf{M} \otimes \mathbf{M}^* \rangle^n \cdot \begin{bmatrix} \langle |I_0(0)|^2 \rangle \\ \langle I_0(0)I_1^*(0) \rangle \\ \langle I_1(0)I_0^*(0) \rangle \\ \langle |I_1(0)|^2 \rangle \end{bmatrix}, \tag{D.18}$$

where

$$\langle \mathbf{M} \otimes \mathbf{M}^* \rangle = \begin{bmatrix} \langle |m_{11}|^2 \rangle & \langle m_{11}m_{12}^* \rangle & \langle m_{12}m_{11}^* \rangle & \langle |m_{12}|^2 \rangle \\ \langle m_{11}m_{21}^* \rangle & \langle m_{11}m_{22}^* \rangle & \langle m_{12}m_{21}^* \rangle & \langle m_{12}m_{22}^* \rangle \\ \langle m_{21}m_{11}^* \rangle & \langle m_{21}m_{12}^* \rangle & \langle m_{22}m_{11}^* \rangle & \langle m_{22}m_{12}^* \rangle \\ \langle |m_{21}|^2 \rangle & \langle m_{21}m_{22}^* \rangle & \langle m_{22}m_{21}^* \rangle & \langle |m_{22}|^2 \rangle \end{bmatrix}. \tag{D.19}$$

The input vector on the right-hand side of Equation (D.18) will normally be simplified. For example, if the signal mode I_0 has deterministic unit input and the spurious mode I_1 has zero input, the input vector will be

$$\langle I(0) \otimes I^*(0) \rangle = I(0) \otimes I^*(0) = \begin{bmatrix} 1 \\ 0 \\ 0 \\ 0 \end{bmatrix}. \tag{D.20}$$

Higher order statistics are obtained in a similar manner. The matrices in the analysis increase in size as the order and/or the number of modes increases.

D.2. MARKOV MATRICES

D.2.1. Markov Chains

Let the random variables and matrices $M(k)$ and $\mathbf{M}(k)$, in Equations (D.1) and (D.9), respectively, each form a Markov chain. In the independent case of Section D.1, these random variables were continuous; in the present Markov case they are restricted to be discrete. For each k, $M(k)$ takes on one of L scalar values

$$a_1, a_2, \cdots, a_L, \tag{D.21}$$

or $\mathbf{M}(k)$ takes on one of L matrix values

$$\mathbf{A}_1, \mathbf{A}_2, \cdots, \mathbf{A}_L. \tag{D.22}$$

We may treat scalars by specializing the matrix case to 1×1 matrices.

Define the unconditional and (conditional) transition probabilities as [4]

$$p_i(k) = P\{\mathbf{M}(k) = \mathbf{A}_i\}, \tag{D.23}$$

$$p_{ij} = P\{\mathbf{M}(k) = \mathbf{A}_i | \mathbf{M}(k-1) = \mathbf{A}_j\}. \tag{D.24}$$

Note that the unconditional probability is in general a function of k, while we assume the transition probability is independent of k. Define the associated column vector and transition matrix as follows:

$$\mathbf{p}^T(k) = [p_1(k) \quad p_2(k) \quad \cdots \quad p_L(k)]. \tag{D.25}$$

$$\mathbf{P} = \begin{bmatrix} p_{11} & p_{12} & \cdots & p_{1L} \\ p_{21} & p_{22} & \cdots & p_{2L} \\ \vdots & \vdots & \ddots & \vdots \\ p_{L1} & p_{L2} & \cdots & p_{LL} \end{bmatrix}. \tag{D.26}$$

Then,

$$\mathbf{p}(k) = \mathbf{P} \cdot \mathbf{p}(k-1). \tag{D.27}$$

184 EXPECTED VALUES OF MATRIX PRODUCTS

We restrict the present treatment to the stationary case:

$$\mathbf{p}(k) = \mathbf{p}. \tag{D.28}$$

From Equations (D.27) and (D.28), the solution for the unit eigenvalue of

$$(\mathbf{P} - \mathcal{I}) \cdot \mathbf{p} = \mathbf{0}, \tag{D.29}$$

where $\mathcal{I}$ is the $L \times L$ unit matrix and $\mathbf{0}$ is the zero $L \times 1$ column vector, yields the unconditional probabilities for the stationary case.

D.2.2. Scalar Variables

We shall require the spectrum of random square-wave coupling in Chapter 8. Let the scalar $M(k)$ denote the coupling in the kth section of the coupling function, n sections long. Then, we must determine $\langle M(n)M(1)\rangle$ in order to calculate the spectrum of the coupling function. We do so below for the Markov case.

Assume that $M(k)$ form a stationary Markov chain, with possible values Equation (D.21), transition matrix Equation (D.26), and stationary unconditional probability vector Equations (D.25) and (D.29). We require the following additional matrices:

$$\mathbf{a} = \begin{bmatrix} a_1 \\ a_2 \\ \vdots \\ a_L \end{bmatrix}; \quad \mathbf{a}_D = \begin{bmatrix} a_1 & 0 & \cdots & 0 \\ 0 & a_2 & \cdots & 0 \\ \vdots & \vdots & \ddots & \vdots \\ 0 & 0 & \cdots & a_L \end{bmatrix}. \tag{D.30}$$

$$\mathbf{p} = \begin{bmatrix} p_1 \\ p_2 \\ \vdots \\ p_L \end{bmatrix}; \quad \mathbf{p}_D = \begin{bmatrix} p_1 & 0 & \cdots & 0 \\ 0 & p_2 & \cdots & 0 \\ \vdots & \vdots & \ddots & \vdots \\ 0 & 0 & \cdots & p_L \end{bmatrix}. \tag{D.31}$$

$$\mathbf{1}^T = \underbrace{\begin{bmatrix} 1 & 1 & \cdots & 1 \end{bmatrix}}_{L \text{ elements}}. \tag{D.32}$$

The joint probability for $M(n)$ and $M(1)$ is

$$P\{M(n) = a_i, M(1) = a_j\}$$
$$= \sum_{k_1 \cdots k_{n-2}} P_{ik_1} P_{k_1 k_2} \cdots P_{k_{n-3} k_{n-2}} P_{k_{n-2} j} P_j. \quad \text{(D.33)}$$

Therefore,

$$\langle M(n)M(1)\rangle = \sum_{ij} a_i P\{M(n) = a_i, M(1) = a_j\} a_j. \quad \text{(D.34)}$$

In matrix notation, Equations (D.33) and (D.34) yield

$$\langle M(n)M(1)\rangle = \mathbf{1}^T \cdot \mathbf{a}_D \cdot \mathbf{P}^{n-1} \cdot \mathbf{p}_D \cdot \mathbf{a}_D \cdot \mathbf{1}, \quad \text{(D.35)}$$

where the component matrices are given by Equations (D.25) and (D.29), (D.26), and (D.30)–(D.32).

Consider the symmetric binary case as an example; i.e., $M(k) = \pm a$ with p the probability that adjacent $M(k)$ are different:

$$\mathbf{P} = \begin{bmatrix} 1-p & p \\ p & 1-p \end{bmatrix}; \quad \mathbf{p} = \begin{bmatrix} \frac{1}{2} \\ \frac{1}{2} \end{bmatrix}; \quad \mathbf{a} = \begin{bmatrix} +a \\ -a \end{bmatrix}. \quad \text{(D.36)}$$

Diagonalizing $\mathbf{P}$:

$$\mathbf{P} = \begin{bmatrix} 1 & 1 \\ -1 & 1 \end{bmatrix} \cdot \begin{bmatrix} 1-2p & 0 \\ 0 & 1 \end{bmatrix} \cdot \begin{bmatrix} 1 & 1 \\ -1 & 1 \end{bmatrix}^{-1};$$
$$\begin{bmatrix} 1 & 1 \\ -1 & 1 \end{bmatrix}^{-1} = \frac{1}{2}\begin{bmatrix} 1 & -1 \\ 1 & 1 \end{bmatrix}. \quad \text{(D.37)}$$

Then Equation (D.35) becomes

$$\langle M(n)M(1)\rangle = \frac{1}{2}\begin{bmatrix} 1 & 1 \end{bmatrix} \cdot \begin{bmatrix} a & 0 \\ 0 & -a \end{bmatrix} \cdot \begin{bmatrix} 1 & 1 \\ -1 & 1 \end{bmatrix}$$
$$\cdot \begin{bmatrix} 1-2p & 0 \\ 0 & 1 \end{bmatrix}^{n-1} \cdot \begin{bmatrix} 1 & -1 \\ 1 & 1 \end{bmatrix}$$
$$\cdot \begin{bmatrix} \frac{1}{2} & 0 \\ 0 & \frac{1}{2} \end{bmatrix} \cdot \begin{bmatrix} a & 0 \\ 0 & -a \end{bmatrix} \cdot \begin{bmatrix} 1 \\ 1 \end{bmatrix}$$
$$= a^2(1-2p)^{n-1}. \quad \text{(D.38)}$$

D.2.3. Markov Matrix Products

Assume the $\mathbf{M}(k)$ of Equation (D.9) form a stationary Markov chain with possible values Equation (D.22), transition matrix Equation (D.26), and stationary unconditional probability vector Equations (D.25) and (D.29). For N modes, $\mathbf{M}(k)$ and $\mathbf{A}_i$ are $N \times N$ matrices. We seek corresponding results to those of Equations (D.12)–(D.20) for independent matrices. Calculation of $\langle \mathbf{M}(n) \cdot \mathbf{M}(n-1) \cdots \mathbf{M}(2) \cdot \mathbf{M}(1) \rangle$ requires the following additional matrices [5]; the size of each matrix is indicated by subscripts:

$$\mathcal{I}_{N \times N} = \begin{bmatrix} 1 & 0 & \cdots & 0 \\ 0 & 1 & \cdots & 0 \\ \vdots & \vdots & \ddots & \vdots \\ 0 & 0 & 0 & 1 \end{bmatrix}. \tag{D.39}$$

$$\mathfrak{P}_{LN \times LN} = \mathbf{P}_{L \times L} \otimes \mathcal{I}_{N \times N} = \begin{bmatrix} p_{11}\mathcal{I} & p_{12}\mathcal{I} & \cdots & p_{1L}\mathcal{I} \\ p_{21}\mathcal{I} & p_{22}\mathcal{I} & \cdots & p_{2L}\mathcal{I} \\ \vdots & \vdots & \ddots & \vdots \\ p_{L1}\mathcal{I} & p_{L2}\mathcal{I} & \cdots & p_{LL}\mathcal{I} \end{bmatrix}. \tag{D.40}$$

$$\mathfrak{A}_{LN \times LN} = \begin{bmatrix} \mathbf{A}_1 & 0 & \cdots & 0 \\ 0 & \mathbf{A}_2 & \cdots & 0 \\ \vdots & \vdots & \ddots & \vdots \\ 0 & 0 & \cdots & \mathbf{A}_L \end{bmatrix}. \tag{D.41}$$

$$\mathfrak{p}_{LN \times N} = \mathbf{p}_{L \times 1} \otimes \mathcal{I}_{N \times N} = \begin{bmatrix} p_1 \mathcal{I} \\ p_2 \mathcal{I} \\ \vdots \\ p_L \mathcal{I} \end{bmatrix}. \tag{D.42}$$

$$\mathfrak{J}_{N \times LN} = [\mathcal{I} \quad \mathcal{I} \quad \cdots \quad \mathcal{I}]. \tag{D.43}$$

The joint probability for the n matrices of Equation (D.9) is

$$P\{\mathbf{M}(n) = \mathbf{A}_i, \mathbf{M}(n-1) = \mathbf{A}_{k_1}, \cdots, \mathbf{M}(2) = \mathbf{A}_{k_{n-2}}, \mathbf{M}(1) = \mathbf{A}_j\}$$
$$= p_{ik_1} p_{k_1 k_2} \cdots p_{k_{n-2} j} p_j. \tag{D.44}$$

D.2. MARKOV MATRICES

Therefore the expected value of the matrix product in Equation (D.9) is

$$\langle \mathbf{M}(n) \cdot \mathbf{M}(n-1) \cdots \mathbf{M}(2) \cdot \mathbf{M}(1) \rangle$$
$$= \sum_{ij} \left(\sum_{k_1 \cdots k_{n-2}} \mathbf{A}_i P_{ik_1} \cdot \mathbf{A}_{k_1} P_{k_1 k_2} \cdots \mathbf{A}_{k_{n-3}} P_{k_{n-3} k_{n-2}} \cdot \mathbf{A}_{k_{n-2}} P_{k_{n-2} j} P_j \right) \cdot \mathbf{A}_j P_j. \tag{D.45}$$

Now

$$(\mathfrak{A} \cdot \mathfrak{B})_{pq} = \mathbf{A}_p P_{pq} \tag{D.46}$$

is the pqth submatrix of $\mathfrak{A} \cdot \mathfrak{B}$. Consequently the quantity in ( ) inside the first summation on the right side of Equation (D.45) is the ijth submatrix of the $(n-1)$th power of $\mathfrak{A} \cdot \mathfrak{B}$:

$$\left(\sum_{k_1 \cdots k_{n-2}} \mathbf{A}_i P_{ik_1} \cdot \mathbf{A}_{k_1} P_{k_1 k_2} \cdots \mathbf{A}_{k_{n-3}} P_{k_{n-3} k_{n-2}} \cdot \mathbf{A}_{k_{n-2}} P_{k_{n-2} j} P_j \right) = (\mathfrak{A} \cdot \mathfrak{B})_{ij}^{(n-1)}. \tag{D.47}$$

Similarly,

$$(\mathfrak{A} \cdot \mathfrak{p})_j = \mathbf{A}_j P_j \tag{D.48}$$

is the jth submatrix of $\mathfrak{A} \cdot \mathfrak{p}$. Therefore Equations (D.45)–(D.48) yield

$$\langle \mathbf{M}(n) \cdot \mathbf{M}(n-1) \cdots \mathbf{M}(2) \cdot \mathbf{M}(1) \rangle$$
$$= \sum_i \sum_j (\mathfrak{A} \cdot \mathfrak{B})_{ij}^{(n-1)} (\mathfrak{A} \cdot \mathfrak{p})_j$$
$$= \sum_i [(\mathfrak{A} \cdot \mathfrak{B})^{(n-1)} (\mathfrak{A} \cdot \mathfrak{p})]_i$$
$$= \mathfrak{J}(\mathfrak{A} \cdot \mathfrak{B})^{(n-1)} (\mathfrak{A} \cdot \mathfrak{p}). \tag{D.49}$$

For the N-mode guide described in Section D.1, substitute Equation (D.49) into the expected value of Equation (D.9) to obtain

$$\langle I(n) \rangle = \mathfrak{J} \cdot (\mathfrak{A} \cdot \mathfrak{B})^{(n-1)} \cdot (\mathfrak{A} \cdot \mathfrak{p}) \cdot \langle I(0) \rangle, \tag{D.50}$$

where $I(0)$ is the vector input, and the other matrix quantities are given in Equations (D.39)–(D.43).

Second- (and higher) order statistics are again found by use of Kronecker products. For second-order, make the following substitutions in the above analysis:

$$\begin{aligned} \mathbf{M}(k) &\longrightarrow \mathbf{M}(k) \otimes \mathbf{M}^*(k) \\ \mathbf{A}_i &\longrightarrow \mathbf{A}_i \otimes \mathbf{A}_i^* \\ N &\longrightarrow N^2 \\ \mathfrak{A} &\longrightarrow \mathfrak{A}^\otimes \end{aligned} \qquad (D.51)$$

Equations (D.39)–(D.43) become

$$\mathcal{I}_{N^2 \times N^2} = \begin{bmatrix} 1 & 0 & \cdots & 0 \\ 0 & 1 & \cdots & 0 \\ \vdots & \vdots & \ddots & \vdots \\ 0 & 0 & 0 & 1 \end{bmatrix}. \qquad (D.52)$$

$$\begin{aligned} \mathfrak{P}_{LN^2 \times LN^2} &= \mathbf{P}_{L \times L} \otimes \mathcal{I}_{N^2 \times N^2} \\ &= \begin{bmatrix} p_{11}\mathcal{I} & p_{12}\mathcal{I} & \cdots & p_{1L}\mathcal{I} \\ p_{21}\mathcal{I} & p_{22}\mathcal{I} & \cdots & p_{2L}\mathcal{I} \\ \vdots & \vdots & \ddots & \vdots \\ p_{L1}\mathcal{I} & p_{L2}\mathcal{I} & \cdots & p_{LL}\mathcal{I} \end{bmatrix}. \end{aligned} \qquad (D.53)$$

$$\mathfrak{A}^\otimes_{LN^2 \times LN^2} = \begin{bmatrix} \mathbf{A}_1 \otimes \mathbf{A}_1^* & 0 & 0 & 0 \\ 0 & \mathbf{A}_2 \otimes \mathbf{A}_2^* & \cdots & 0 \\ \vdots & \vdots & \ddots & \vdots \\ 0 & 0 & \cdots & \mathbf{A}_L \otimes \mathbf{A}_L^* \end{bmatrix}. \qquad (D.54)$$

$$\mathfrak{p}_{LN^2 \times N^2} = \mathbf{P}_{L \times 1} \otimes \mathcal{I}_{N^2 \times N^2} = \begin{bmatrix} p_1 \mathcal{I} \\ p_2 \mathcal{I} \\ \vdots \\ p_L \mathcal{I} \end{bmatrix}. \qquad (D.55)$$

$$\mathfrak{J}_{N^2 \times LN^2} = \underbrace{[\mathcal{I}_{N^2 \times N^2} \quad \mathcal{I}_{N^2 \times N^2} \quad \cdots \quad \mathcal{I}_{N^2 \times N^2}]}_{L \text{ unit matrices}}. \quad (D.56)$$

Note that $\mathfrak{A}^{\otimes}_{LN^2 \times LN^2}$ of Equation (D.54) is *not* equal to the Kronecker product of $\mathfrak{A}_{LN \times LN}$ of Equation (D.41) with itself. Then the expected value of Equation (D.15) yields

$$\langle I(n) \otimes I^*(n) \rangle = \mathfrak{J} \cdot (\mathfrak{A}^{\otimes} \cdot \mathfrak{P})^{(n-1)} \cdot (\mathfrak{A}^{\otimes} \cdot \mathfrak{p}) \cdot \langle I(0) \otimes I^*(0) \rangle, \quad (D.57)$$

where the matrix quantities are now given in Equations (D.52)–(D.56). If the input consists of only a single deterministic (signal) mode, the input vector $I(0) \otimes I^*(0)$ will contain only a single nonzero element.

We illustrate these results for the symmetric binary case; here the transition matrix **P** and unconditional probability vector **p** are given by Equation (D.36). Then, Equations (D.50) and (D.57) become

$$\langle I(n) \rangle = \frac{1}{2} [\mathcal{I} \quad \mathcal{I}] \cdot \begin{bmatrix} (1-p)\mathbf{A}_1 & p\mathbf{A}_1 \\ p\mathbf{A}_2 & (1-p)\mathbf{A}_2 \end{bmatrix}^{(n-1)} \cdot \begin{bmatrix} \mathbf{A}_1 \\ \mathbf{A}_2 \end{bmatrix} \cdot \langle I(0) \rangle, \quad (D.58)$$

where $\mathcal{I}$ is the 2×2 unit matrix, and

$$\langle I(n) \otimes I^*(n) \rangle$$
$$= \frac{1}{2} [\mathcal{I} \quad \mathcal{I}] \cdot \begin{bmatrix} (1-p)\mathbf{A}_1 \otimes \mathbf{A}_1^* & p\mathbf{A}_1 \otimes \mathbf{A}_1^* \\ p\mathbf{A}_2 \otimes \mathbf{A}_2^* & (1-p)\mathbf{A}_2 \otimes \mathbf{A}_2^* \end{bmatrix}^{(n-1)}$$
$$\cdot \begin{bmatrix} \mathbf{A}_1 \otimes \mathbf{A}_1^* \\ \mathbf{A}_2 \otimes \mathbf{A}_2^* \end{bmatrix} \cdot \langle I(0) \otimes I^*(0) \rangle, \quad (D.59)$$

where $\mathcal{I}$ is the 4×4 unit matrix. For $p = \frac{1}{2}$, the matrices $\mathbf{M}(k)$ are independent; Equations (D.58) and (D.59) reduce to Equations (D.14) and (D.17), respectively.

REFERENCES

1. R. Bellman, *Introduction to Matrix Analysis*, McGraw-Hill, New York, 1970; Chapters 12 and 15.
2. Harrison E. Rowe and D. T. Young, "Transmission Distortion in Multimode Random Waveguides," *IEEE Transactions on Microwave Theory and Techniques*, Vol. MTT-20, June 1972, pp. 349–365.
3. Harrison E. Rowe, "Waves with Random Coupling and Random Propagation Constants," *Applied Scientific Research*, Vol. 41, 1984, pp. 237–255.
4. Athanasios Papoulis, *Probability, Random Variables, and Stochastic Processes*, McGraw-Hill, New York, 2nd. ed., 1984; Chapter 12.
5. Behram Homi Bharucha, *On the Stability of Randomly Varying Systems*, Ph.D. Thesis, University of California, 1961; Appendix D.

APPENDIX E

Time- and Frequency-Domain Statistics

E.1. SECOND-ORDER IMPULSE RESPONSE STATISTICS

We derive the time-domain results used in Section 3.7 for a wide-sense stationary random transfer function [1]. The filter impulse response and transfer function of a real, causal filter are given in Equations (3.98)–(3.101):

$$g(t) = \int_{-\infty}^{\infty} \mathcal{G}(f)e^{j2\pi ft} df. \qquad (E.1)$$

$$g(t) = g^*(t); \qquad g(t) = 0, \quad t < 0. \qquad (E.2)$$

$$\mathcal{G}(f) = \int_0^{\infty} g(t)e^{-j2\pi ft} dt. \qquad (E.3)$$

$$\mathcal{G}(f) = \mathcal{G}^*(-f). \qquad (E.4)$$

We denote the Fourier transform relation as follows:

$$g(t) \Longleftrightarrow \mathcal{G}(f). \qquad (E.5)$$

The transfer function is wide-sense stationary by Equations (3.102)–(3.103):

$$\mathcal{R}(\nu) = \langle \mathcal{G}(f+\nu)\mathcal{G}^*(f) \rangle; \qquad \mathcal{R}(\nu) = \mathcal{R}^*(-\nu). \qquad (E.6)$$

$$\langle \mathcal{G}(f) \rangle = \langle \mathcal{G} \rangle. \qquad (E.7)$$

We assume that $\mathcal{G}(f)$ has no periodic components; then

$$\mathcal{R}(\infty) = \mathcal{R}(-\infty) = \langle \mathcal{G} \rangle^2. \tag{E.8}$$

It is convenient to divide $\mathcal{G}(f)$ into d.c. and a.c. components. Toward this end denote averages over frequency by $\overline{}$. Then,

$$\overline{\mathcal{G}} = \lim_{F \to \infty} \frac{1}{2F} \int_{-F}^{F} \mathcal{G}(f) df; \qquad \mathcal{G}(f) = \overline{\mathcal{G}} + \mathcal{G}_{ac}(f). \tag{E.9}$$

We show that $\overline{\mathcal{G}}$ is deterministic; i.e.,

$$\overline{\mathcal{G}} = \langle \mathcal{G} \rangle. \tag{E.10}$$

Thus, from Equations (E.8) and (E.9),

$$\langle (\overline{\mathcal{G}} - \langle \mathcal{G} \rangle)^2 \rangle = \lim_{F \to \infty} \frac{1}{(2F)^2} \int_{-F}^{F} \int_{-F}^{F} \langle \mathcal{G}(f')\mathcal{G}(f'') \rangle df' df''$$

$$- 2\overline{\langle \mathcal{G} \rangle} \langle \mathcal{G} \rangle + \langle \mathcal{G} \rangle^2$$

$$= \lim_{F \to \infty} \frac{1}{(2F)^2} \int_{-F}^{F} \int_{-F}^{F} \mathcal{R}(f' - f'') df' df'' - \langle \mathcal{G} \rangle^2$$

$$= \mathcal{R}(\infty) - \langle \mathcal{G} \rangle^2 = 0. \tag{E.11}$$

Therefore,

$$\mathcal{G}(f) = \langle \mathcal{G} \rangle + \mathcal{G}_{ac}(f); \qquad \langle \mathcal{G}_{ac}(f) \rangle = 0. \tag{E.12}$$

$$\mathcal{R}(\nu) = \langle \mathcal{G} \rangle^2 + \mathcal{R}_{ac}(\nu); \qquad \mathcal{R}_{ac}(\nu) = \langle \mathcal{G}_{ac}(f + \nu) \mathcal{G}_{ac}^*(f) \rangle. \tag{E.13}$$

From Equation (3.104),

$$\mathcal{P}(t) = \int_{-\infty}^{\infty} \mathcal{R}(\nu) e^{j2\pi t\nu} d\nu, \qquad \mathcal{P}(t) = \mathcal{P}^*(t), \tag{E.14}$$

where $\mathcal{P}(-t)$ is the transfer function spectral density. Substituting Equation (E.13):

$$\mathcal{P}(t) = \langle \mathcal{G} \rangle^2 \delta(t) + \mathcal{P}_{ac}(t); \qquad \mathcal{P}_{ac}(t) = \int_{-\infty}^{\infty} \mathcal{R}_{ac}(\nu) e^{-j2\pi t\nu} d\nu. \tag{E.15}$$

$\mathcal{P}_{ac}(t)$ has no δ-function components.

E.1. SECOND-ORDER IMPULSE RESPONSE STATISTICS

Additional properties of the real and imaginary parts of $\mathcal{G}(f)$ and $\mathcal{R}(\nu)$, not of direct interest here, are given in [1].

Cascade $\mathcal{G}(f)$ with an ideal band-pass filter, to yield the overall transfer function $\mathcal{H}(f)$ of Equation (3.105):

$$\mathcal{H}(f) = \begin{cases} \mathcal{G}(f), & |f - f_0| < B, \\ 0, & |f - f_0| > B, \end{cases} \quad B < f_0. \tag{E.16}$$

The envelope of the overall response is given in terms of the impulse response $h(t)$ and its Hilbert transform $\widehat{h}(t)$ by Equation (3.106),

$$\nu_h(t) = |h(t) + j\widehat{h}(t)|, \tag{E.17}$$

where $\widehat{h}(t)$ is given by Equation (3.107) in terms of the convolution operator $\star$:

$$\widehat{h}(t) = \frac{1}{\pi t} \star h(t) = \frac{1}{\pi} \int_{-\infty}^{\infty} \frac{h(\tau)}{t - \tau} d\tau. \tag{E.18}$$

The positive and negative frequency content of $h(t)$ are

$$h_+(t) = \frac{1}{2}[h(t) + j\widehat{h}(t)] \iff \mathcal{H}_+(f) = \frac{1}{2}(1 + \text{sgn } f)\mathcal{H}(f), \tag{E.19}$$

$$h_-(t) = \frac{1}{2}[h(t) - j\widehat{h}(t)] \iff \mathcal{H}_-(f) = \frac{1}{2}(1 - \text{sgn } f)\mathcal{H}(f), \tag{E.20}$$

where sgn( ) is the signum function:

$$\text{sgn } f = \begin{cases} 1, & f > 0. \\ -1, & f < 0. \end{cases} \tag{E.21}$$

Figure E.1 shows a typical band-pass spectrum.

Then, using Equation (E.16),

$$\nu_h^2(t) \iff 4\mathcal{H}_+(f) \star \mathcal{H}_-(f) = 4\mathcal{H}_+(f) \star \mathcal{H}_+^*(f)$$

$$= 4 \int_{f_0-B-(f-|f|)/2}^{f_0+B-(f+|f|)/2} \mathcal{G}(a+f)\mathcal{G}^*(a) da. \tag{E.22}$$

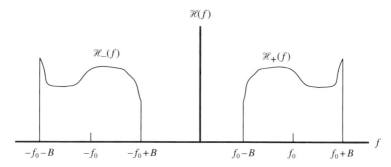

FIGURE E.1. Band-pass spectrum.

Substituting Equation (E.6),

$$\langle \nu_h^2(t) \rangle = 8B \int_{-2B}^{2B} \left(1 - \frac{|f|}{2B}\right) \mathcal{R}(f) e^{j2\pi ft} df. \tag{E.23}$$

From Equation (E.14),

$$\langle \nu_h^2(t) \rangle = 8B \left\{ 2B \left(\frac{\sin 2\pi Bt}{2\pi Bt} \right)^2 \right\} \star \mathcal{P}(t). \tag{E.24}$$

From Equation (E.15),

$$\langle \nu_h^2(t) \rangle = 8B \langle \mathcal{G} \rangle^2 \left\{ 2B \left(\frac{\sin 2\pi Bt}{2\pi Bt} \right)^2 \right\}$$
$$+ 8B \left\{ 2B \left(\frac{\sin 2\pi Bt}{2\pi Bt} \right)^2 \right\} \star \mathcal{P}_{ac}(t). \tag{E.25}$$

Since

$$\lim_{B \to \infty} \left\{ 2B \left(\frac{\sin 2\pi Bt}{2\pi Bt} \right)^2 \right\} = \delta(t), \tag{E.26}$$

we have the result stated in Equation (3.108):

$$\langle \nu_h^2(t) \rangle \approx 8B \cdot \mathcal{P}(t), \qquad B \text{ large}, \tag{E.27}$$

where $1/(2B)$ is small compared to the width of $\mathcal{P}_{ac}(t)$, or equivalently B is large compared to the correlation bandwidth of $\mathcal{G}(f)$.

Finally, we show that the response of a wide-sense stationary transfer function to white noise is linear in intensity [2]. From Equation (3.110)

$$r_y^2(t) = 4y_+(t)y_-(t)$$
$$= \int_{-\infty}^{\infty}\int_{-\infty}^{\infty} \mathcal{I}(\tau_1)\mathcal{I}(\tau_2)x(\tau_1)x(\tau_2)$$
$$\cdot 4h_+(t-\tau_1)h_-(t-\tau_2)d\tau_1 d\tau_2. \qquad (E.28)$$

Averaging over the input and using Equation (3.109),

$$\langle r_y^2(t)\rangle_x = \int_{-\infty}^{\infty} \mathcal{I}^2(\tau)r_h^2(t-\tau)d\tau. \qquad (E.29)$$

Averaging over the transfer function,

$$\langle r_y^2(t)\rangle = \langle\langle r_y^2(t)\rangle_x\rangle_h$$
$$= \int_{-\infty}^{\infty} \mathcal{I}^2(\tau)\langle r_h^2(t-\tau)\rangle d\tau$$
$$= \mathcal{I}^2(t) \star \langle r_h^2(t)\rangle, \qquad (E.30)$$

as given in Equation (3.111).

E.2. TIME-DOMAIN ANALYSIS

We derive the impulse response of the idealized two-mode guide of Section 3.2 directly from the coupled line equations.

Consider a length of line dx for two coupled modes described by Equations (3.4)–(3.5); the coupling coefficient for this section is

$$jc(x)dx = j\,\mathrm{sgn}\,f \cdot C_0 d(x)\,dx, \qquad (E.31)$$

from Equations (3.15) and (3.18). The corresponding impulse response is the inverse Fourier transform of this quantity:

$$C_0 d(x)dx \cdot j\,\mathrm{sgn}\,f \iff -C_0 d(x)dx\frac{1}{\pi t}. \qquad (E.32)$$

196 TIME- AND FREQUENCY-DOMAIN STATISTICS

Thus, a signal $\delta(t)$ at position x in one mode excites a response $-C_0 d(x) dx (1/\pi t)$ at x in the other guide; similarly, a signal $-1/\pi t$ at x in one mode excites a response $C_0 d(x) dx \delta(t)$ at x in the other mode.

Assume a unit impulse incident on the signal mode at the guide input $z = 0$, and apply the method of successive approximations in the time domain. The elementary partial signal-mode impulse response that has made n transitions from signal to spurious mode at $x_{2n}, x_{2n-2}, \cdots, x_4, x_2$, and n transitions from spurious to signal mode at $x_{2n-1}, x_{2n-3}, \cdots, x_3, x_1$, is given by

$$\delta\left(t - \frac{x_1 - x_2 + \cdots + x_{2n-1} - x_{2n}}{L} \Delta T\right) \cdot e^{\Delta\alpha(x_1 - x_2 + \cdots + x_{2n-1} - x_{2n})}$$
$$\cdot C_0^{2n} d(x_1) d(x_2) \cdots d(x_{2n-1}) d(x_{2n}) dx_1 dx_2 \cdots dx_{2n-1} dx_{2n},$$
$$L > x_1 > x_2 > \cdots > x_{2n-1} > x_{2n}, \quad \text{(E.33)}$$

where ΔT is given by Equation (3.20). Integrating Equation (E.33) over the variables $\{x_i\}$, $1 \leq i \leq 2n$, yields the partial impulse response composed of all components that have made exactly $2n$ transitions; note that this quantity normalized and Equation (B.2) are Fourier transforms, as in Equations (3.16) and (3.17). Finally, summing over n, $0 \leq n \leq \infty$ yields the total signal mode impulse response $g_{\Delta\alpha}(t)$ of Equations (3.22)–(3.23). The properties of Appendix B follow immediately.

A time-domain model that contained some of the essential features of Equation (E.33) gave the first quantitative understanding of the impulse response of random guides [2].

REFERENCES

1. Harrison E. Rowe and D. T. Young, "Transmission Distortion in Multimode Random Waveguides," *IEEE Transactions on Microwave Theory and Techniques*, Vol. MTT-20, June 1972, pp. 349–365.
2. S. D. Personick, "Time Dispersion in Dielectric Waveguides," *Bell System Technical Journal*, Vol. 50, March 1971, pp. 843–859.

APPENDIX F

Symmetric Slab Waveguide—Lossless TE Modes

F.1. GENERAL RESULTS

Figure F.1 shows a symmetric dielectric slab waveguide of refractive index n_1 and thickness d suspended in a medium of index n_2.

Both media are nonmagnetic; i.e., they have the same permeability as free space. We summarize the properties of this guide as follows [1]:

$$\lambda_0 = \frac{1}{f\sqrt{\mu_0 \epsilon_0}}, \quad \text{free-space wavelength.} \tag{F.1}$$

$$k_0 = \frac{2\pi}{\lambda_0}, \quad \text{propagation constant in free space.} \tag{F.2}$$

$$k_i = \frac{2\pi}{\lambda_i}, \quad \text{propagation constant in medium of index } n_i. \tag{F.3}$$

$$n_i = \sqrt{\frac{\epsilon_i}{\epsilon_0}} = \frac{\lambda_0}{\lambda_i}; \quad n_1 > n_2. \tag{F.4}$$

The propagation constants of TE modes traveling in the z direction, indexed by the subscript ν, are given by

$$\beta_\nu = \sqrt{n_1^2 k_0^2 - \kappa_\nu^2} = \sqrt{k_1^2 - \kappa_\nu^2}, \tag{F.5}$$

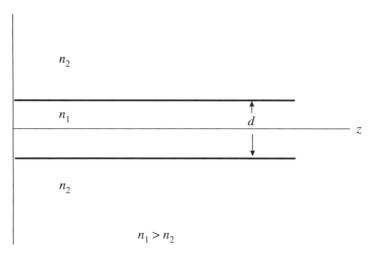

FIGURE F.1. Ideal symmetric slab waveguide.

where

$$\gamma = \sqrt{(n_1^2 - n_2^2)k_0^2 - \kappa^2} = \sqrt{k_1^2 - k_2^2 - \kappa^2} \tag{F.6}$$

and κ_ν is one of the solutions of the eigenvalue equations:

$$\tan \frac{\kappa d}{2} = \frac{\gamma}{\kappa}, \quad \text{even modes,} \tag{F.7}$$

$$\tan \frac{\kappa d}{2} = -\frac{\kappa}{\gamma}, \quad \text{odd modes.} \tag{F.8}$$

κ_ν must be positive and γ real; hence,

$$0 < \kappa_\nu < \kappa_{\max} = \sqrt{n_1^2 - n_2^2} \cdot k_0 = \sqrt{k_1^2 - k_2^2}. \tag{F.9}$$

Equation (F.9) implies

$$k_2 < \beta_\nu < k_1; \tag{F.10}$$

the propagation constants of the TE modes in the waveguide lie between the propagation constants of the core and cladding media.

F.1. GENERAL RESULTS

It is helpful to rewrite Equation (F.6) in terms of κ_{max} of Equation (F.9):

$$\gamma = \sqrt{\kappa_{max}^2 - \kappa^2}. \tag{F.11}$$

Equations (F.7) and (F.8) may be solved with Equation (F.11) for the eigenvalues by plotting $\tan \kappa d/2$, γ/κ, and $-\kappa/\gamma$ on the same graph and finding the intersections. Figure F.2 shows such a plot for a special case of interest in Sections 4.5, 7.1, and F.2.

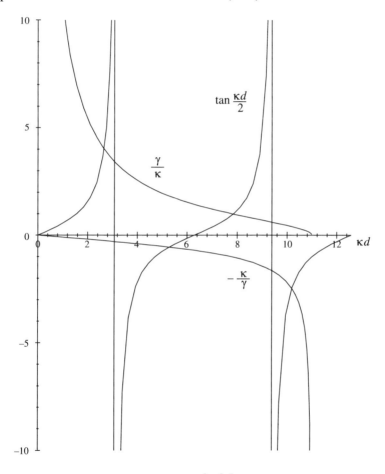

$\kappa_{max} d = 3.5\pi$

FIGURE F.2. Solution for the eigenvalues.

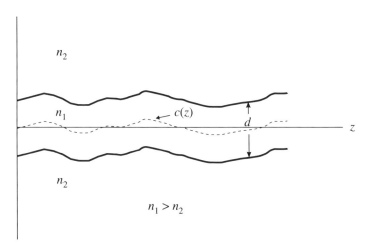

FIGURE F.3. Symmetric slab waveguide with random straightness deviation.

For this case, $\kappa_{max}d = 3.5\pi$, there are four TE modes, two even and two odd. More generally, for $(N-1)\pi < \kappa_{max}d < N\pi$ there are N TE modes, $\lfloor(N+1)/2\rfloor$ even and $\lfloor N/2\rfloor$ odd, where $\lfloor x \rfloor$ represents the greatest integer contained in x.

Finally, consider the slab guide with random straightness deviation $c(z)$, shown in Figure F.3. The coupling coefficients between two forward-traveling TE modes are given as follows:

$$C_{\mu\nu} = -\frac{2\kappa_\mu \kappa_\nu \varepsilon_{\mu\nu}}{\left[\beta_\mu \beta_\nu \left(d + \frac{1}{\gamma_\mu}\right)\left(d + \frac{1}{\gamma_\nu}\right)\right]^{1/2}}. \quad (F.12)$$

$$\varepsilon_{\mu\nu} = \begin{cases} 1, & \text{one mode even and the other odd.} \\ 0, & \text{both modes even or both odd.} \end{cases} \quad (F.13)$$

$C_{\mu\nu}$ is to be substituted into Equations (3.35) and (3.13); $c(z)$ in Equation (3.9) represents the straightness deviation of the random guide, shown in Figure F.3.

F.2. EXAMPLE

We take the following parameters for the waveguide of Figures F.1 and F.3:

$$\lambda_0 = 10^{-6} \text{ m.}; \quad n_1 = 1.01; \quad n_2 = 1. \quad (F.14)$$

From Equation (F.9),

$$\kappa_{max} = \sqrt{1.01^2 - 1} \frac{2\pi}{10^{-6}} \text{ m}^{-1}. \tag{F.15}$$

Take

$$\kappa_{max} d = 3.5\pi \tag{F.16}$$

as in Figure F.2. Then, from Equations (F.15) and (F.16),

$$d = \frac{3.5 \times 10^{-6}}{2\sqrt{1.01^2 - 1}} = 1.2343 \times 10^{-5} \text{ m}. \tag{F.17}$$

Substituting Equation (F.16) into Equation (F.11),

$$\frac{\gamma}{\kappa} = \sqrt{\frac{(3.5\pi)^2}{(\kappa d)^2} - 1}. \tag{F.18}$$

Equation (F.18) with Equations (F.7) and (F.8) yield the following solutions for the eigenvalues, illustrated in Figure F.2[1]:

$$\begin{aligned}\kappa_0 &= 2.1501 \times 10^5 \text{ m}^{-1}, & \kappa_1 &= 4.2785 \times 10^5 \text{ m}^{-1}.\\ \kappa_2 &= 6.3496 \times 10^5 \text{ m}^{-1}, & \kappa_3 &= 8.2581 \times 10^5 \text{ m}^{-1}.\end{aligned} \tag{F.19}$$

Substituting these values into Equations (F.5), (F.6), and (F.12), the mode propagation constants and coupling coefficients to be substituted into Equations (3.35) and (3.36) are

$$\mathbf{\Gamma} = j \times 10^6 \begin{bmatrix} 6.3424 & 0 & 0 & 0 \\ 0 & 6.3316 & 0 & 0 \\ 0 & 0 & 6.3142 & 0 \\ 0 & 0 & 0 & 6.2921 \end{bmatrix} \text{ m}^{-1}. \tag{F.20}$$

---

[1] We number the eigenvalues $0, 1, 2, 3, \cdots$, to correspond with the notation used throughout the text.

$$\mathbf{C} = 10^9 \begin{bmatrix} 0 & 2.3225 & 0 & 4.4966 \\ 2.3225 & 0 & 6.8737 & 0 \\ 0 & 6.8737 & 0 & 13.3083 \\ 4.4966 & 0 & 13.3083 & 0 \end{bmatrix} \text{m}^{-2}. \quad \text{(F.21)}$$

REFERENCE

1. Dietrich Marcuse, *Theory of Dielectric Optical Waveguides*, 2nd ed., Academic Press, New York, 1991.

APPENDIX G

Equal Propagation Constants

We treat completely degenerate guides, in which all modes have identical propagation constants, by elementary methods. These results corroborate the general results obtained by Kronecker product methods in Chapters 3, 4, 6, and 7. The present degenerate case is without physical significance.

Consider the two-mode case with zero loss and constant propagation parameters:

$$\Gamma_0 = \Gamma_1 = j\beta. \tag{G.1}$$

Assume white coupling, Equations (3.1)–(3.3), with δ-function covariance:

$$\begin{aligned} R_c(\zeta) &= \langle c(z+\zeta)c(z)\rangle = S_0\delta(\zeta). \\ S(\nu) &= \int_{-\infty}^{\infty} R_c(\zeta)e^{-j2\pi\nu\zeta}d\zeta = S_0. \end{aligned} \tag{G.2}$$

Assume unit coherent single-mode input:

$$I_0(0) = 1, \qquad I_1(0) = 0. \tag{G.3}$$

From Equations (2.7) and (2.8),

$$\begin{aligned} I_0(z) &= e^{-j\beta z}\cos[C(z)]. \\ I_1(z) &= e^{-j\beta z}j\sin[C(z)]. \end{aligned} \tag{G.4}$$

204 EQUAL PROPAGATION CONSTANTS

$C(z)$ is the integrated coupling coefficient:

$$C(z) = \int_0^z c(x)dx. \tag{G.5}$$

For white coupling, $C(z)$ is Gaussian with zero mean

$$\langle C(z) \rangle = 0, \tag{G.6}$$

and variance

$$\langle C^2(z) \rangle = \int_0^z \int_0^z \langle c(x)c(y) \rangle dx\, dy$$
$$= S_0 \int_0^z dy \int_0^z \delta(x-y)dx = S_0 z. \tag{G.7}$$

The following results are obtained by using the characteristic function of a Gaussian random variable to evaluate the trigonometric expected values:

$$\begin{aligned}\mathcal{I}_0(z) &= \langle I_0(z) \rangle = e^{-j\beta z} \langle \cos[C(z)] \rangle = e^{-j\beta z} e^{-\frac{1}{2}S_0 z}. \\ \mathcal{I}_1(z) &= \langle I_1(z) \rangle = e^{-j\beta z} \langle \sin[C(z)] \rangle = 0. \end{aligned} \tag{G.8}$$

$$\begin{aligned}\mathcal{P}_0(z) &= \langle P_0(z) \rangle = \langle |I_0(z)|^2 \rangle = \langle \cos^2[C(z)] \rangle = \frac{1}{2}\left[1 + e^{-2S_0 z}\right]. \\ \mathcal{P}_1(z) &= \langle P_1(z) \rangle = \langle |I_1(z)|^2 \rangle = \langle \sin^2[C(z)] \rangle = \frac{1}{2}\left[1 - e^{-2S_0 z}\right]. \end{aligned} \tag{G.9}$$

$$\begin{aligned}\mathcal{Q}_0(z) &= \langle P_0^2(z) \rangle = \langle |I_0(z)|^4 \rangle = \langle \cos^4[C(z)] \rangle \\ &= \frac{3}{8} + \frac{1}{8}e^{-8S_0 z} + \frac{1}{2}e^{-2S_0 z}. \\ \mathcal{Q}_4(z) &= \langle P_1^2(z) \rangle = \langle |I_0(z)|^4 \rangle = \langle \sin^4[C(z)] \rangle \\ &= \frac{3}{8} + \frac{1}{8}e^{-8S_0 z} - \frac{1}{2}e^{-2S_0 z}. \end{aligned} \tag{G.10}$$

Equations (G.8) and (G.9) agree with Equations (4.23) and (4.26). Note that the former were obtained here for equal propagation constants; the latter are more general, valid for arbitrary propagation constants. Equation (G.10) agrees with Equation (4.50).

EQUAL PROPAGATION CONSTANTS

Next, assume two-mode coherent input, replacing Equation (G.3) by

$$I_0(0) = 1, \quad I_1(0) = 1. \tag{G.11}$$

Equations (2.7) and (2.8) yield

$$I_0(z) = I_1(z) = e^{-j\beta z} e^{jC(z)}, \tag{G.12}$$

where $C(z)$ is again given by Equation (G.5). Then

$$\mathcal{I}_0(z) = \mathcal{I}_1(z) = e^{-j\beta z} \langle e^{jC(z)} \rangle = e^{-j\beta z} e^{-\frac{1}{2}S_0 z}. \tag{G.13}$$

Equation (G.13) agrees with Equation (4.31). Here

$$P_0(z) = |I_0(z)|^2 = 1, \quad P_1(z) = |I_1(z)|^2 = 1; \tag{G.14}$$

the mode powers (but *not* the complex mode amplitudes) are deterministic in the degenerate case, and consequently have zero variance, in agreement with the results of Section 4.3.2 for the $\Delta\beta = 0$ case.

For two-mode incoherent input,

$$I_0(0) = e^{j\theta_0}, \quad I_1(0) = e^{j\theta_1}, \tag{G.15}$$

with θ_0 and θ_1 independent random variables, uniformly distributed from 0 to 2π. Equations (2.7) and (2.8) become

$$I_0(z) = e^{-j\beta z} \left\{ e^{j\theta_0} \cos[C(z)] + j e^{j\theta_1} \sin[C(z)] \right\}, \tag{G.16}$$

with a corresponding result for I_1, obtained by interchanging the subscripts 0 and 1. Then,

$$\begin{aligned} \mathcal{Q}_0(z) = \mathcal{Q}_4(z) &= \langle P_0^2(z) \rangle = \langle |I_0(z)|^4 \rangle \\ &= \langle \{1 + \sin[2C(z)] \sin(\theta_0 - \theta_1)\}^2 \rangle \\ &= \frac{5}{4} - \frac{1}{4} e^{-8S_0 z}. \end{aligned} \tag{G.17}$$

Equations (G.17) and (4.40) substituted into Equation (4.48) confirm the result of Equation (4.54).

206 EQUAL PROPAGATION CONSTANTS

The N-mode degenerate case may be treated in a similar direct manner. For constant propagation parameter Equation (2.51) becomes

$$\mathbf{\Gamma}(z) = \Gamma \mathcal{I}, \tag{G.18}$$

where Γ is the common propagation constant of all N modes,

$$\Gamma = \alpha + j\beta, \tag{G.19}$$

with $\mathcal{I}$ the unit matrix of Equation (2.52). Then from Equation (2.53)

$$\mathbf{I}(z) = e^{-\Gamma z} e^{jC(z)\mathbf{C}} \cdot \mathbf{I}(0). \tag{G.20}$$

$\mathbf{C}$ is the coupling coefficient matrix of Equation (2.45) and $C(z)$ is the integrated coupling coefficient of Equation (G.5). Then using the notation of Equations (3.38), (3.57), and (3.70):

$$\mathcal{I}(z) = \langle \mathbf{I}(z) \rangle = e^{-\Gamma z} \langle e^{jC(z)\mathbf{C}} \rangle \cdot \mathcal{I}(0). \tag{G.21}$$

$$\mathcal{P}(z) = \langle \mathbf{I}(z) \otimes \mathbf{I}^*(z) \rangle = e^{-2\alpha z} \langle e^{jC(z)\mathbf{C}} \otimes e^{-jC(z)\mathbf{C}} \rangle \cdot \mathcal{P}(0). \tag{G.22}$$

$$\mathcal{P}_4(z) = \langle \mathbf{I}(z) \otimes \mathbf{I}^*(z) \otimes \mathbf{I}(z) \otimes \mathbf{I}^*(z) \rangle$$
$$= e^{-4\alpha z} \langle e^{jC(z)\mathbf{C}} \otimes e^{-jC(z)\mathbf{C}} \otimes e^{jC(z)\mathbf{C}} \otimes e^{-jC(z)\mathbf{C}} \rangle \cdot \mathcal{P}_4(0). \tag{G.23}$$

The general results of Chapter 3 for arbitrary propagation constants specialize to the degenerate case by substituting Equation (G.18) into Equations (3.41) and (3.61), to yield the following solutions:

$$\mathcal{I}(z) = e^{-\Gamma z} e^{-\frac{1}{2}S_0 z \mathbf{C}^2} \cdot \mathcal{I}(0). \tag{G.24}$$

$$\mathcal{P}(z) = e^{-2\alpha z} e^{-\frac{1}{2}S_0 z (\mathcal{I} \otimes \mathbf{C}^2 + \mathbf{C}^2 \otimes \mathcal{I} - 2\mathbf{C} \otimes \mathbf{C})} \cdot \mathcal{P}(0). \tag{G.25}$$

We omit the corresponding result obtained from Equation (3.72), for brevity. Equations (G.24) and (G.25) must be identical to Equations (G.21) and (G.22), respectively; therefore,

$$\langle e^{jC(z)\mathbf{C}} \rangle = e^{-\frac{1}{2}S_0 z \mathbf{C}^2}. \tag{G.26}$$

$$\langle e^{jC(z)\mathbf{C}} \otimes e^{-jC(z)\mathbf{C}} \rangle = e^{-\frac{1}{2}S_0 z (\mathcal{I} \otimes \mathbf{C}^2 + \mathbf{C}^2 \otimes \mathcal{I} - 2\mathbf{C} \otimes \mathbf{C})}. \tag{G.27}$$

Equations (G.26) and (G.27) may be verified by series expansion of the matrix exponentials and use of the Gaussian property of $C(z)$ to obtain its moments. We perform this calculation for Equation (G.26), where the second moment of $C(z)$ is given by Equation (G.7) for white $c(z)$:

$$\langle e^{jC(z)\mathbf{C}} \rangle = \left\langle \sum_{n=0}^{\infty} \frac{[jC(z)\mathbf{C}]^n}{n!} \right\rangle = \sum_{n=0}^{\infty} \frac{(-1)^n \langle C^{2n}(z) \rangle}{(2n)!} \mathbf{C}^{2n}$$

$$= \sum_{n=0}^{\infty} \frac{\left(-\frac{1}{2}S_0 z \mathbf{C}^2\right)^n}{n!} = e^{-\frac{1}{2}S_0 z \mathbf{C}^2}. \qquad (G.28)$$

Similar treatments verify Equation (G.27), and show the exponent of the expected value in Equation (G.23) to be identical to the expression in [ ] on the right-hand side of Equation (3.72) with the substitution of Equation (G.18).

Finally, the above results for white coupling may be extended to completely degenerate guides with general coupling spectra $S(\nu)$ and covariance $R_c(\zeta)$ as follows:

$$R_c(\zeta) = \langle c(z+\zeta)c(z) \rangle, \qquad S(\nu) = \int_{-\infty}^{\infty} R_c(\zeta) e^{-j2\pi\nu\zeta} d\zeta. \qquad (G.29)$$

Assume that $S(\nu)$ is low pass, or has a low-pass component; i.e.,

$$S(0) \neq 0. \qquad (G.30)$$

Let ν_c represent the bandwidth of the coupling and λ_c its correlation distance:

$$\lambda_c = \frac{1}{\nu_c}. \qquad (G.31)$$

Most of the power of coupling $c(z)$ is contained in the band $|\nu| < \nu_c$; $c(z)$ and $c(z+\zeta)$ are essentially uncorrelated for $\zeta \gg \lambda_c$. The integrated coupling coefficient of Equation (G.5) has zero mean, Equation (G.6). For its variance, Equation (G.7) is replaced by

$$\langle C^2(z) \rangle = \int_0^z \int_0^z R_c(x-y) dx\, dy. \qquad (G.32)$$

Calculation of this quantity requires complete knowledge of the covariance; if the coupling $c(z)$ is Gaussian, so is the integrated coupling $C(z)$, and its variance determines all of its higher moments, as in the prior results for white coupling. For simplicity, the following treatment is restricted to the large z case. Then, Equation (G.32) yields

$$\langle C^2(z) \rangle \approx S(0)z, \qquad z \gg \lambda_c. \tag{G.33}$$

In this case, $C(z)$ will be approximately Gaussian, whatever the statistics of $c(z)$. The prior results of the present appendix, for completely degenerate guides with white coupling, apply to general coupling spectra by replacing the white spectral density S_0 of Equation (G.2) by the low-frequency spectral density $S(0)$ of Equation (G.29), if z is sufficiently large.

APPENDIX H

Asymptotic Form of Coupled Power Equations

Consider $\mathbf{M}_1$ of Equations (4.104)–(4.105):

$$\mathbf{M}_1 = \mathbf{V} \otimes \mathbf{V}^*. \tag{H.1}$$

$$\mathbf{V} = e^{\Gamma z} \cdot \mathbf{C} \cdot e^{-\Gamma z}, \qquad \mathbf{V}^* = e^{\Gamma^* z} \cdot \mathbf{C} \cdot e^{-\Gamma^* z}. \tag{H.2}$$

Recalling the notation of Equation (2.45) or (3.35), we have the following results for the matrix elements:

$$\left.\begin{aligned}
(\mathbf{V})_{ii} &= C_{i-1,i-1} = 0 \\
(\mathbf{V})_{ij} &= e^{\Delta\Gamma_{i-1,j-1}z} C_{i-1,j-1} = e^{(\Delta\alpha_{i-1,j-1}+j\Delta\beta_{i-1,j-1})z} C_{i-1,j-1} \\
(\mathbf{V}^*)_{ii} &= C_{i-1,i-1} = 0 \\
(\mathbf{V}^*)_{ij} &= e^{\Delta\Gamma^*_{i-1,j-1}z} C_{i-1,j-1} = e^{(\Delta\alpha_{i-1,j-1}-j\Delta\beta_{i-1,j-1})z} C_{i-1,j-1}
\end{aligned}\right\},$$

$$1 \leq i, j \leq N. \tag{H.3}$$

Now the submatrix $\mathbf{M}_{1|ij}$ defined by Equation (4.108) is given by

$$\mathbf{M}_{1|ij} = (\mathbf{V})_{ij}\mathbf{V}^* = e^{\Delta\Gamma_{i-1,j-1}z} C_{i-1,j-1} \mathbf{V}^*, \quad 1 \leq i, j \leq N. \tag{H.4}$$

We find the asymptotic results by setting to zero each element of Equation (H.4) that contains an exponential factor. Denote the elements of $\mathbf{M}_{1|ij}$ by $(\mathbf{M}_{1|ij})_{k\ell}$; then assuming no degenerate modes, only the following elements survive:

$$(\mathbf{M}_{1|ij})_{ij} = e^{2\Delta\alpha_{i-1,j-1}z} C^2_{i-1,j-1}, \quad 1 \leq i, j \leq N; \quad \text{(H.5)}$$

this expression is zero for $i = j$. The submatrices $\mathbf{M}_{1|ii}$ on the main diagonal of $\mathbf{M}_1$ are zero; the submatrices $\mathbf{M}_{1|ij}$, $i \neq j$ off of the main diagonal of $\mathbf{M}_1$ have only a single nonzero element, in the ith row and jth column of $\mathbf{M}_{1|ij}$.

The asymptotic form of $\mathbf{M}_2$ and $\mathbf{M}_3$ of Equations (4.104)–(4.105) is found directly; only the diagonal terms survive. By using Equation (4.82), the diagonal submatrices have only the following nonzero elements in the asymptotic case:

$$(\mathbf{M}_{2|ii})_{kk} = (\mathbf{C}^2)_{kk} = \sum_{\ell=0}^{N-1} C^2_{k-1,\ell}, \quad 1 \leq i, k \leq N. \quad \text{(H.6)}$$

$$(\mathbf{M}_{3|ii})_{kk} = (\mathbf{C}^2)_{ii} = \sum_{\ell=0}^{N-1} C^2_{i-1,\ell}, \quad 1 \leq i, k \leq N. \quad \text{(H.7)}$$

The $\ell = k - 1$ and $\ell = i - 1$ terms in the respective summations of these two relations are zero.

The results of Equations (H.5)–(H.7) are substituted into Equation (4.109).

APPENDIX I

Differential Equations Corresponding to Matrix Equations

I.1. SCALAR CASE

Consider the difference equation

$$y[k\Delta z] = My[(k-1)\Delta z]. \tag{I.1}$$

M and Δz are fixed parameters. The solution to Equation (I.1) is

$$y(k\Delta z) = M^k y(0). \tag{I.2}$$

We seek an equivalent differential equation:

$$\frac{dy}{dz} = Dy. \tag{I.3}$$

The solution to Equation (I.3), evaluated at the points $z = k\Delta z$, is

$$y(k\Delta z) = e^{Dk\Delta z} y(0). \tag{I.4}$$

Equations (I.2) and (I.4) will be identical if

$$D = \frac{\ln M}{\Delta z}. \tag{I.5}$$

In the perturbation case,

$$M \approx 1 + E\Delta z, \qquad |E|\Delta z \ll 1. \tag{I.6}$$

Here Equation (I.5) becomes

$$D \approx E. \tag{I.7}$$

I.2. MATRIX CASE

Consider the vector–matrix difference equation

$$\mathbf{y}[k\Delta z] = \mathbf{M} \cdot \mathbf{y}[(k-1)\Delta z], \tag{I.8}$$

where $\mathbf{y}$ is a column vector, $\mathbf{M}$ a fixed matrix, and Δz a fixed parameter. The solution to Equation (I.8) is

$$\mathbf{y}(k\Delta z) = \mathbf{M}^k \cdot \mathbf{y}(0). \tag{I.9}$$

The equivalent vector–matrix differential equation is

$$\frac{d}{dz}\mathbf{y}(z) = \mathbf{D} \cdot \mathbf{y}(z). \tag{I.10}$$

Its solution is

$$\mathbf{y}(k\Delta z) = e^{\mathbf{D}k\Delta z} \cdot \mathbf{y}(0). \tag{I.11}$$

The two solutions agree if

$$e^{\mathbf{D}\Delta z} = \mathbf{M}. \tag{I.12}$$

Diagonalizing

$$\mathbf{M} = \mathbf{L}_M \cdot \mathbf{\Lambda}_M \cdot \mathbf{L}_M^{-1}, \qquad \mathbf{\Lambda}_M = \begin{bmatrix} \lambda_{M1} & 0 & \cdots & 0 \\ 0 & \lambda_{M2} & \cdots & 0 \\ \vdots & \vdots & \ddots & \vdots \\ 0 & 0 & \cdots & \lambda_{MN} \end{bmatrix}. \tag{I.13}$$

I.2. MATRIX CASE

$$\mathbf{D} = \mathbf{L}_D \cdot \mathbf{\Lambda}_D \cdot \mathbf{L}_D^{-1}, \qquad \mathbf{\Lambda}_D = \begin{bmatrix} \lambda_{D1} & 0 & \cdots & 0 \\ 0 & \lambda_{D2} & \cdots & 0 \\ \vdots & \vdots & \ddots & \vdots \\ 0 & 0 & \cdots & \lambda_{DN} \end{bmatrix}. \qquad (I.14)$$

Then,

$$e^{\mathbf{D}\Delta z} = \mathbf{L}_D \cdot \begin{bmatrix} e^{\lambda_{D1}\Delta z} & 0 & \cdots & 0 \\ 0 & e^{\lambda_{D2}\Delta z} & \cdots & 0 \\ \vdots & \vdots & \ddots & \vdots \\ 0 & 0 & \cdots & e^{\lambda_{DN}\Delta z} \end{bmatrix} \cdot \mathbf{L}_D^{-1}. \qquad (I.15)$$

Equation (I.12) requires that

$$\mathbf{L}_D = \mathbf{L}_M; \qquad \lambda_{Di} = \frac{\ln \lambda_{Mi}}{\Delta z}, \quad i = 1 \cdots N. \qquad (I.16)$$

Simplifying notation, diagonalize the matrix $\mathbf{M}$ of the difference equation, Equation (I.8):

$$\mathbf{M} = \mathbf{L} \cdot \begin{bmatrix} \lambda_1 & 0 & \cdots & 0 \\ 0 & \lambda_2 & \cdots & 0 \\ \vdots & \vdots & \ddots & \vdots \\ 0 & 0 & \cdots & \lambda_N \end{bmatrix} \cdot \mathbf{L}^{-1}. \qquad (I.17)$$

Then, the equivalent differential equation is given by Equation (I.10) with [1]

$$\mathbf{D} = \mathbf{L} \cdot \begin{bmatrix} \frac{\ln \lambda_1}{\Delta z} & 0 & \cdots & 0 \\ 0 & \frac{\ln \lambda_2}{\Delta z} & \cdots & 0 \\ \vdots & \vdots & \ddots & \vdots \\ 0 & 0 & \cdots & \frac{\ln \lambda_N}{\Delta z} \end{bmatrix} \cdot \mathbf{L}^{-1}. \qquad (I.18)$$

For perturbation theory,

$$\mathbf{M} \approx \mathcal{I} + \mathbf{E}\Delta z, \qquad \|\mathbf{E}\|\Delta z \ll 1, \tag{I.19}$$

where $\mathcal{I}$ is the $N \times N$ unit matrix and $\|\ \|$ denotes the matrix norm, as in Equation (A.25). Here Equation (I.18) becomes

$$\mathbf{D} \approx \mathbf{E}. \tag{I.20}$$

REFERENCE

1. Neil A. Jackman, *Limitations and Tolerances in Optical Devices*, Appendix C, Ph.D. Thesis, Stevens Institute of Technology, Castle Point, Hoboken, NJ, 1994.

APPENDIX J

Random Square-Wave Coupling Statistics

J.1. INTRODUCTION

The coupling function of Equation (8.1) and Figure 8.1 may be written as follows:

$$c(z) = \sum c_k \operatorname{rect}\left(\frac{z}{\ell} - \frac{1}{2} - k\right), \tag{J.1}$$

where

$$\operatorname{rect}\left(\frac{z}{\ell} - \frac{1}{2} - k\right) = \begin{cases} 1, & z = k\ell < z < (k+1)\ell. \\ 0, & \text{otherwise}. \end{cases} \tag{J.2}$$

The fundamental rect( ) function has Fourier transform expressed in terms of the sinc function as follows:

$$\int_{-\infty}^{\infty} \operatorname{rect}\left(\frac{z}{\ell} - \frac{1}{2}\right) \cdot e^{-j2\pi\nu z} dz = e^{-j\pi\nu\ell} \ell \operatorname{sinc} \nu\ell$$

$$= e^{-j\pi\nu\ell} \ell \frac{\sin \pi\nu\ell}{\pi\nu\ell}. \tag{J.3}$$

The sequence $\{c_k\}$ has been called a discrete function [1] or a discrete-time function [2]. Assume that $\{c_k\}$ is stationary, with discrete covariance

$$\phi_c(\kappa) = \langle c_{k+\kappa} c_k \rangle. \tag{J.4}$$

We assume the c_k have zero mean:

$$\langle c_k \rangle = 0. \tag{J.5}$$

Then $S(\nu)$, the spectral density of the coupling $c(z)$ defined in Equations (3.2) or (6.2), is given by

$$S(\nu) = \ell \left(\frac{\sin \pi \nu \ell}{\pi \nu \ell}\right)^2 \sum_{\kappa=-\infty}^{\infty} \phi_c(\kappa) e^{-j2\pi\nu\kappa\ell}. \tag{J.6}$$

Two special cases are of interest.
For independent $\{c_k\}$,

$$\phi_c(0) = \langle c_k^2 \rangle \equiv \langle c^2 \rangle; \qquad \phi_c(\kappa) = 0, \quad \kappa \neq 0. \tag{J.7}$$

Equation (J.6) becomes

$$S(\nu) = \langle c^2 \rangle \ell \left(\frac{\sin \pi \nu \ell}{\pi \nu \ell}\right)^2. \tag{J.8}$$

Finally, consider a sequence $\{c_k\}$ that is a weighted average of a stationary, independent, zero-mean, binary sequence $\{d \cdot d_k\}$:

$$c_k = d a_k, \qquad a_k = \sum_{n=-\infty}^{\infty} d_n h(k-n) = \sum_{n=-\infty}^{\infty} h(n) d_{k-n}, \tag{J.9}$$

where $h(\ )$ is the weight function,

$$d_k = \pm 1, \qquad \langle d_k \rangle = 0, \tag{J.10}$$

and the covariance of d_k is given as follows:

$$\phi_d(0) = \langle d_k^2 \rangle = 1; \qquad \phi_d(\kappa) = \langle d_{k+\kappa} d_k \rangle = 0, \quad \kappa \neq 0. \tag{J.11}$$

The autocorrelation of the weight function is defined as

$$\varphi_h(\kappa) = \sum_{k=-\infty}^{\infty} h(k+\kappa) h(k). \tag{J.12}$$

Then $\{c_k\}$ has covariance

$$\phi_c(\kappa) = d^2\varphi_h(\kappa), \tag{J.13}$$

to be substituted in Equation (J.6).

The following two weight functions generate multi-level, pseudo-Gaussian, square-wave coupling coefficients $c(z)$, with low-pass and band-pass coupling spectra $S(\nu)$, respectively:

$$h_{\text{LP}}(k) = \begin{cases} 1, & |k| \leq K \\ 0, & |k| > K. \end{cases} \tag{J.14}$$

$$h_{\text{BP}}(k) = \begin{cases} -1, & -K \leq k \leq -1. \\ 0, & k = 0. \\ 1, & 1 \leq k \leq K. \end{cases} \tag{J.15}$$

$c(z)$ has $2K+1$ or $2K$ equally spaced levels, for weight functions $h_{\text{LP}}(k)$ and $h_{\text{BP}}(k)$, respectively. The corresponding autocorrelation functions, for substitution in Equation (J.13), are

$$\varphi_{\text{LP}}(\kappa) = \begin{cases} 2K+1-|\kappa|, & |\kappa| \leq 2K+1. \\ 0 & |\kappa| \geq 2K+1. \end{cases} \tag{J.16}$$

$$\varphi_{\text{BP}}(\kappa) = \begin{cases} 2K, & \kappa = 0. \\ 2K - 3|\kappa| + 1, & 1 \leq |\kappa| \leq K. \\ -2K + |\kappa| - 1, & K+1 \leq |\kappa| \leq 2K+1. \\ 0, & 2K+1 \leq |\kappa|. \end{cases} \tag{J.17}$$

J.2. BINARY SECTIONS

Let the c_k of Equations (8.1) and (J.1) take on the values $\{a_i\}$, $1 \leq i \leq L$, with unconditional probabilities

$$p_i(k) = p_i = P\{c_k = a_i\}, \tag{J.18}$$

independent of k for stationary $\{c_k\}$. The joint probabilities are

$$P\{c_k = a_i, \, c_n = a_j\}, \quad k \neq n. \tag{J.19}$$

218 RANDOM SQUARE-WAVE COUPLING STATISTICS

Denote the transition probabilities for adjacent values of k by

$$p_{ij} = P\{c_k = a_i | c_{k-1} = a_j\} = \frac{P\{c_{k-1} = a_j, c_k = a_i\}}{P\{c_{k-1} = a_j\}}, \qquad (J.20)$$

again independent of k. We assume the symmetric binary case, $L = 2$, in the following two subsections:

$$a_1 = a, \quad a_2 = -a; \qquad p_1 = p_2 = \frac{1}{2}. \qquad (J.21)$$

J.2.1. Independent

For independent $\{c_k\}$,

$$P\{c_k = a_i, c_n = a_j\} = p_i p_j; \qquad p_{ij} = p_i. \qquad (J.22)$$

The covariance, Equation (J.7), becomes

$$\phi_c(0) = a^2; \qquad \phi_c(\kappa) = 0, \quad \kappa \neq 0. \qquad (J.23)$$

The coupling spectrum, Equation (J.8), is

$$S(\nu) = a^2 \ell \left(\frac{\sin \pi \nu \ell}{\pi \nu \ell} \right)^2. \qquad (J.24)$$

J.2.2. Markov

Let p be the probability that adjacent c_k are different:

$$p = P\{c_{k-1} \neq c_k\}, \qquad 1 - p = P\{c_{k-1} = c_k\}. \qquad (J.25)$$

The transition matrix for adjacent c_k is given by **P** of Equation (D.36); the covariance of Equation (J.4) is obtained from Equation (D.38) as

$$\phi_c(\kappa) = a^2 (1 - 2p)^{|\kappa|}. \qquad (J.26)$$

By substituting Equation (J.26) into Equation (J.6), the coupling spectrum is

$$S(\nu) = a^2 \ell \left(\frac{\sin \pi \nu \ell}{\pi \nu \ell} \right)^2 \frac{1 - (1 - 2p)^2}{1 + (1 - 2p)^2 - 2(1 - 2p) \cos 2\pi \nu \ell}. \qquad (J.27)$$

For $p = \frac{1}{2}$, these results specialize to those of the preceding section for the independent case.

J.3. MULTI-LEVEL MARKOV SECTIONS

J.3.1. Low-Pass—Six Levels

Consider a six-level, low-pass c_k described by Equations (J.9)–(J.14). Setting $K = 2$ in Equation (J.16):

$$\begin{array}{c|cccccc} |\kappa| & 0 & 1 & 2 & 3 & 4 & \geq 5 \\ \hline \varphi_{LP}(\kappa) & 5 & 4 & 3 & 2 & 1 & 0 \end{array}. \quad (J.28)$$

Substituting in Equation (J.13), the coupling spectrum Equation (J.6) becomes

$$S(\nu) = d^2 \ell \left(\frac{\sin \pi \nu \ell}{\pi \nu \ell} \right)^2 \left[5 + 2 \sum_{\kappa=1}^{4} (5 - \kappa) \cos(2\pi\nu\kappa\ell) \right]. \quad (J.29)$$

We omit the closed form for the summation in this equation, subsequently using MAPLE to evaluate it directly.

The $\{c_k\}$ for this model are correlated, pseudo-Gaussian random variables, with the following probability distribution:

$$c_k = d a_k; \quad \begin{array}{c|cccccc} a_k & -5 & -3 & -1 & 1 & 3 & 5 \\ \hline P\{a_k\} & \frac{1}{32} & \frac{5}{32} & \frac{10}{32} & \frac{10}{32} & \frac{5}{32} & \frac{1}{32} \end{array}. \quad (J.30)$$

We need to describe this example as a Markov process to make use of the results of Section D.2. From Equations (J.9) and (J.14),

$$c_k = d a_k, \quad a_k = d_{k-2} + d_{k-1} + d_k + d_{k+1} + d_{k+2}. \quad (J.31)$$

We denote the state of the system at time k by the notation $STATE(k)$, characterized by the five-dimensional vector

220 RANDOM SQUARE-WAVE COUPLING STATISTICS

$[d_{k-2}\ d_{k-1}\ d_k\ d_{k+1}\ d_{k+2}]$. The state can take on the (vector) values S_i, $1 \leq i \leq 32$, given in the following table:

i	S_i					$a_{i\text{LP}}$
1	−1	−1	−1	−1	−1	−5
2	−1	−1	−1	−1	1	−3
3	−1	−1	−1	1	−1	−3
4	−1	−1	−1	1	1	−1
5	−1	−1	1	−1	−1	−3
6	−1	−1	1	−1	1	−1
7	−1	−1	1	1	−1	−1
8	−1	−1	1	1	1	1
9	−1	1	−1	−1	−1	−3
10	−1	1	−1	−1	1	−1
11	−1	1	−1	1	−1	−1
12	−1	1	−1	1	1	1
13	−1	1	1	−1	−1	−1
14	−1	1	1	−1	1	1
15	−1	1	1	1	−1	1
16	−1	1	1	1	1	3
17	1	−1	−1	−1	−1	−3
18	1	−1	−1	−1	1	−1
19	1	−1	−1	1	−1	−1
20	1	−1	−1	1	1	1
21	1	−1	1	−1	−1	−1
22	1	−1	1	−1	1	1
23	1	−1	1	1	−1	1
24	1	−1	1	1	1	3
25	1	1	−1	−1	−1	−1
26	1	1	−1	−1	1	1
27	1	1	1	1	−1	1
28	1	1	−1	1	1	3
29	1	1	1	−1	−1	1
30	1	1	1	−1	1	3
31	1	1	1	1	−1	3
32	1	1	1	1	1	5

(J.32)

The left column i indexes the states; the middle column lists the state vectors S_i; the final column $a_{i\text{LP}}$, calculated from S_i and Equation (J.31), determines the coupling constant c_k of the kth section as

$$c_k = da_{i\text{LP}}, \qquad STATE(k) = S_i, \qquad (J.33)$$

for the ith state at time k.

We require the matrix quantities $\mathbf{p}_D$, $\mathbf{a}_D$, and $\mathbf{P}$ in Equation (D.35); these three matrices have dimension 32×32 in the present example. The 32 states have equal probability, yielding the diagonal matrix

$$\mathbf{p}_D = \frac{1}{32} \cdot \begin{bmatrix} 1 & & & 0 \\ & 1 & & \\ & & \ddots & \\ 0 & & & 1 \end{bmatrix}. \tag{J.34}$$

The diagonal matrix $\mathbf{a}_D$ of Equation (D.35) is found from the right-hand column of Equations (J.32) and (J.33), expressed as four 16×16 submatrices as follows:

$$\mathbf{a}_D = d \left[\begin{array}{c|c} \mathbf{a}_{D11} & 0 \\ \hline 0 & \mathbf{a}_{D22} \end{array} \right]. \tag{J.35}$$

$\mathbf{0}$ represents the zero submatrix, and the diagonal submatrices are

$$\mathbf{a}_{D11} = \begin{bmatrix} -5 & & & & & & & & & & & & & & & 0 \\ & -3 & & & & & & & & & & & & & & \\ & & -3 & & & & & & & & & & & & & \\ & & & -1 & & & & & & & & & & & & \\ & & & & -3 & & & & & & & & & & & \\ & & & & & -1 & & & & & & & & & & \\ & & & & & & -1 & & & & & & & & & \\ & & & & & & & 1 & & & & & & & & \\ & & & & & & & & -3 & & & & & & & \\ & & & & & & & & & -1 & & & & & & \\ & & & & & & & & & & -1 & & & & & \\ & & & & & & & & & & & 1 & & & & \\ & & & & & & & & & & & & -1 & & & \\ & & & & & & & & & & & & & 1 & & \\ & & & & & & & & & & & & & & 1 & \\ 0 & & & & & & & & & & & & & & & 3 \end{bmatrix}. \tag{J.36}$$

$$\mathbf{a}_{D22} = \begin{bmatrix} -3 & & & & & & & & & & 0 \\ & -1 & & & & & & & & & \\ & & -1 & & & & & & & & \\ & & & 1 & & & & & & & \\ & & & & -1 & & & & & & \\ & & & & & 1 & & & & & \\ & & & & & & 1 & & & & \\ & & & & & & & 3 & & & \\ & & & & & & & -1 & & & \\ & & & & & & & & 1 & & \\ & & & & & & & & & 1 & \\ & & & & & & & & & & 3 \\ & & & & & & & & & & 1 \\ & & & & & & & & & & 3 \\ & & & & & & & & & & 3 \\ 0 & & & & & & & & & & 5 \end{bmatrix} \quad (\text{J}.37)$$

Finally, the elements $\{p_{ij}\}$ of the matrix $\mathbf{P}$ of Equation (D.26) are

$$p_{ij} = P\{STATE(K) = S_i | STATE(K-1) = S_j\}. \quad (\text{J}.38)$$

These conditional probabilities follow directly from the middle columns of Equation (J.32). For example, suppose that

$$STATE(k-1) = S_9 = [\overset{d_{k-3}}{-1} \quad \overset{d_{k-2}}{1} \quad \overset{d_{k-1}}{-1} \quad \overset{d_k}{-1} \quad \overset{d_{k+1}}{-1}]. \quad (\text{J}.39)$$

Then since d_{k+2} can take on only the values ± 1, $STATE(K)$ can have only the two possible values

$$STATE(k) = S_{17} = [\overset{d_{k-2}}{1} \quad \overset{d_{k-1}}{-1} \quad \overset{d_k}{-1} \quad \overset{d_{k+1}}{-1} \quad \overset{d_{k+2}}{-1}] \quad (\text{J}.40)$$

and

$$STATE(k) = S_{18} = [\overset{d_{k-2}}{1} \quad \overset{d_{k-1}}{-1} \quad \overset{d_k}{-1} \quad \overset{d_{k+1}}{-1} \quad \overset{d_{k+2}}{1}]. \quad (\text{J}.41)$$

J.3. MULTI-LEVEL MARKOV SECTIONS

Therefore only two elements of the ninth column of **P** are nonzero:

$$p_{17,9} = p_{18,9} = \frac{1}{2}; \quad p_{i9} = 0, \quad i \neq 17 \text{ or } 18. \quad (J.42)$$

P is given as two submatrices as follows:

$$\mathbf{P} = \begin{bmatrix} \mathbf{P} \mid \mathbf{P} \end{bmatrix}, \quad (J.43)$$

where

$$\mathbf{P} = \frac{1}{2}\begin{bmatrix}
1 & & & & & & & & & & & & & & & & 0 \\
1 & & & & & & & & & & & & & & & & \\
& 1 & & & & & & & & & & & & & & & \\
& 1 & & & & & & & & & & & & & & & \\
& & 1 & & & & & & & & & & & & & & \\
& & 1 & & & & & & & & & & & & & & \\
& & & 1 & & & & & & & & & & & & & \\
& & & 1 & & & & & & & & & & & & & \\
& & & & 1 & & & & & & & & & & & & \\
& & & & 1 & & & & & & & & & & & & \\
& & & & & 1 & & & & & & & & & & & \\
& & & & & 1 & & & & & & & & & & & \\
& & & & & & 1 & & & & & & & & & & \\
& & & & & & 1 & & & & & & & & & & \\
& & & & & & & 1 & & & & & & & & & \\
& & & & & & & 1 & & & & & & & & & \\
& & & & & & & & 1 & & & & & & & & \\
& & & & & & & & 1 & & & & & & & & \\
& & & & & & & & & 1 & & & & & & & \\
& & & & & & & & & 1 & & & & & & & \\
& & & & & & & & & & 1 & & & & & & \\
& & & & & & & & & & 1 & & & & & & \\
& & & & & & & & & & & 1 & & & & & \\
& & & & & & & & & & & 1 & & & & & \\
& & & & & & & & & & & & 1 & & & & \\
& & & & & & & & & & & & 1 & & & & \\
& & & & & & & & & & & & & 1 & & & \\
& & & & & & & & & & & & & 1 & & & \\
& & & & & & & & & & & & & & 1 & & \\
& & & & & & & & & & & & & & 1 & & \\
0 & & & & & & & & & & & & & & & 1 & \\
& & & & & & & & & & & & & & & & 1
\end{bmatrix}.$$

(J.44)

224 RANDOM SQUARE-WAVE COUPLING STATISTICS

The coupling covariance is given by Equations (J.13), (J.16), and (J.28); the same result has alternatively been obtained from the Markov model, i.e., Equations (D.35), (J.34)–(J.37), and (J.44), using MAPLE for the calculations.

J.3.2. Band-Pass—Five Levels

Set $K = 2$ in Equation (J.17) to yield for a five-level, band-pass c_k

$\|\kappa\|$	0	1	2	3	4	≥ 5
$\varphi_{BP}(\kappa)$	4	2	-1	-2	-1	0

(J.45)

From Equations (J.13) and (J.6),

$$S(\nu) = d^2\ell \left(\frac{\sin \pi\nu\ell}{\pi\nu\ell}\right)^2 \{4 + 2[2\cos(2\pi\nu\ell) - \cos(4\pi\nu\ell) - 2\cos(6\pi\nu\ell) - \cos(8\pi\nu\ell)]\}. \quad (J.46)$$

Equations (J.9) and (J.15) yield

$$c_k = da_k, \qquad a_k = -d_{k-2} - d_{k-1} + d_{k+1} + d_{k+2}. \quad (J.47)$$

Equation (J.32) remains valid with the exception of the last column, relabeled a_{iBP}, calculated from the S_i of Equations (J.32) and (J.47):

i	1	2	3	4	5	6	7	8	9	10	11	12	13	14	15	16
a_{iBP}	0	2	2	4	0	2	2	4	-2	0	0	2	-2	0	0	2

i	17	18	19	20	21	22	23	24	25	26	27	28	29	30	31	32
a_{iBP}	-2	0	0	2	-2	0	0	2	-4	-2	-2	0	-4	-2	-2	0

(J.48)

The matrix $\mathbf{p}_D$ remains the same as Equation (J.34). The diagonal elements of $\mathbf{a}_D$ are now given by a_{iBP} of Equation (J.48); Equations

(J.36) and (J.37), for the diagonal submatrices of Equation (J.35), now become

$$\mathbf{a}_{D11} = \begin{bmatrix} 0 & & & & & & & & & & & & & & & 0 \\ & 2 & & & & & & & & & & & & & & \\ & & 2 & & & & & & & & & & & & & \\ & & & 4 & & & & & & & & & & & & \\ & & & & 0 & & & & & & & & & & & \\ & & & & & 2 & & & & & & & & & & \\ & & & & & & 2 & & & & & & & & & \\ & & & & & & & 4 & & & & & & & & \\ & & & & & & & & -2 & & & & & & & \\ & & & & & & & & & 0 & & & & & & \\ & & & & & & & & & & 0 & & & & & \\ & & & & & & & & & & & 2 & & & & \\ & & & & & & & & & & & & -2 & & & \\ & & & & & & & & & & & & & 0 & & \\ & & & & & & & & & & & & & & 0 & \\ 0 & & & & & & & & & & & & & & & 2 \end{bmatrix},$$
(J.49)

$$\mathbf{a}_{D22} = \begin{bmatrix} -2 & & & & & & & & & & & & & & & 0 \\ & 0 & & & & & & & & & & & & & & \\ & & 0 & & & & & & & & & & & & & \\ & & & 2 & & & & & & & & & & & & \\ & & & & -2 & & & & & & & & & & & \\ & & & & & 0 & & & & & & & & & & \\ & & & & & & 0 & & & & & & & & & \\ & & & & & & & 2 & & & & & & & & \\ & & & & & & & & -4 & & & & & & & \\ & & & & & & & & & -2 & & & & & & \\ & & & & & & & & & & -2 & & & & & \\ & & & & & & & & & & & 0 & & & & \\ & & & & & & & & & & & & -4 & & & \\ & & & & & & & & & & & & & -2 & & \\ & & & & & & & & & & & & & & -2 & \\ 0 & & & & & & & & & & & & & & & 0 \end{bmatrix}.$$
(J.50)

Finally, Equations (J.43) and (J.44) remain valid for **P**.

The coupling covariance computed from the Markov model—Equations (D.35), (J.34), (J.35), (J.49), and (J.50)—agrees with that obtained directly from Equations (J.13), (J.17), and (J.45).

REFERENCES

1. H. E. Rowe, *Signals and Noise in Communication Systems*, D. Van Nostrand, New York, 1965.
2. Alan V. Oppenheim and Ronald W. Schafer, *Discrete-Time Signal Processing*, Prentice Hall, Englewood Cliffs, NJ, 1989.

APPENDIX K

Matrix for a Multi-Layer Structure

The electric and magnetic fields for the linearly polarized wave in Figure 9.1 consist of forward and backward waves:

$$E(z) = E^+(z) + E^-(z), \qquad H(z) = H^+(z) + H^-(z), \qquad (K.1)$$

where

$$H^+(z) = \frac{E^+(z)}{Z}, \qquad H^-(z) = -\frac{E^-(z)}{Z}, \qquad (K.2)$$

and Z is the characteristic impedance of the dielectric medium:

$$Z = \sqrt{\frac{\mu_0}{\varepsilon}}. \qquad (K.3)$$

The forward and backward waves have the following z dependence:

$$E^+(z) = E^+ e^{-j\beta z}, \qquad E^-(z) = E^- e^{+j\beta z}, \qquad (K.4)$$

where the propagation constant is

$$\beta = \frac{2\pi n}{\lambda}, \qquad (K.5)$$

λ is the free-space wavelength,

$$\lambda = \frac{1}{f\sqrt{\mu_0 \varepsilon_0}}, \qquad (K.6)$$

228 MATRIX FOR A MULTI-LAYER STRUCTURE

and n is the index of refraction of the dielectric medium,

$$n = \sqrt{\frac{\varepsilon}{\varepsilon_0}}. \tag{K.7}$$

The total electric field and its forward and backward components are related by the following matrix equations:

$$\begin{bmatrix} E(z) \\ Z_0 H(z) \end{bmatrix} = \begin{bmatrix} 1 & 1 \\ n & -n \end{bmatrix} \cdot \begin{bmatrix} E^+(z) \\ E^-(z) \end{bmatrix}. \tag{K.8}$$

$$\begin{bmatrix} E^+(z) \\ E^-(z) \end{bmatrix} = \frac{1}{2} \begin{bmatrix} 1 & \frac{1}{n} \\ 1 & -\frac{1}{n} \end{bmatrix} \cdot \begin{bmatrix} E(z) \\ Z_0 H(z) \end{bmatrix}. \tag{K.9}$$

Here Z_0 is the characteristic impedance of free space:

$$Z_0 = \sqrt{\frac{\mu_0}{\varepsilon_0}}. \tag{K.10}$$

Apply these relations to the kth layer in Figure 9.1. The forward and backward electric fields at z_k and z_{k-1} are related as follows:

$$\begin{bmatrix} E^+(z_{k-1}) \\ E^-(z_{k-1}) \end{bmatrix} = \begin{bmatrix} e^{j\beta_k d_k} & 0 \\ 0 & e^{-j\beta_k d_k} \end{bmatrix} \cdot \begin{bmatrix} E^+(z_k) \\ E^-(z_k) \end{bmatrix}, \tag{K.11}$$

where from Equation (K.5)

$$\beta_k = \frac{2\pi n_k}{\lambda}. \tag{K.12}$$

From Equations (K.8), (K.9), and (K.11),

$$\begin{bmatrix} E(z_{k-1}) \\ Z_0 H(z_{k-1}) \end{bmatrix} = \mathbf{M}_k \cdot \begin{bmatrix} E(z_k) \\ Z_0 H(z_k) \end{bmatrix}, \tag{K.13}$$

where

$$\mathbf{M}_k = \frac{1}{2} \begin{bmatrix} 1 & 1 \\ n & -n \end{bmatrix} \cdot \begin{bmatrix} e^{j\beta_k d_k} & 0 \\ 0 & e^{-j\beta_k d_k} \end{bmatrix} \cdot \begin{bmatrix} 1 & \frac{1}{n} \\ 1 & -\frac{1}{n} \end{bmatrix}. \tag{K.14}$$

MATRIX FOR A MULTI-LAYER STRUCTURE

Evaluating this matrix product,

$$\mathbf{M}_k = \begin{bmatrix} \cos\phi_k & \dfrac{j\sin\phi_k}{n_k} \\ jn_k \sin\phi_k & \cos\phi_k \end{bmatrix}, \tag{K.15}$$

where ϕ_k is the phase shift associated with the kth layer,

$$\phi_k = \beta_k d_k = \frac{2\pi}{\lambda} n_k d_k. \tag{K.16}$$

The electric and magnetic fields are continuous across the boundaries at $0, z_1, \cdots, z_\ell$. Therefore,

$$\begin{bmatrix} E(z_0-) \\ Z_0 H(z_0-) \end{bmatrix} = \prod_{k=1}^{\ell} \mathbf{M}_k \cdot \begin{bmatrix} E(z_\ell+) \\ Z_0 H(z_\ell+) \end{bmatrix}. \tag{K.17}$$

By using Equation (K.9), the input and output quantities in Figure 9.1 are related as follows:

$$\begin{bmatrix} E_i \\ E_r \end{bmatrix} = \frac{1}{2} \begin{bmatrix} 1 & \dfrac{1}{n_i} \\ 1 & -\dfrac{1}{n_i} \end{bmatrix} \cdot \prod_{k=1}^{\ell} \mathbf{M}_k \cdot \begin{bmatrix} 1 & 1 \\ n_t & -n_t \end{bmatrix} \cdot \begin{bmatrix} E_t \\ 0 \end{bmatrix}. \tag{K.18}$$

Index

Average transfer functions:
 random coupling:
  examples:
   exponential covariance, 126
   white spectrum, 52, 56, 58, 69–70, 72–73
  multi-mode:
   almost-white spectra, 106
   general spectra, 112, 118
   white spectrum, 28–29, 67
  nondegenerate, 76
  two modes:
   almost-white spectra, 102–103
   general spectra, 109–110, 116
   white spectrum, 27
  random propagation parameters, 91–92

Coupled line equations, 1, 5, 6
 multi-mode, 15, 22
  approximate solutions, 18–19
  exact solutions, 16–17
  small coupling, 19, 108
 series solutions:
  multi-mode, 167
  two modes, 165–166
 two modes, 6, 22, 87
  approximate solutions, 11–15
  exact solutions, 8–11, 88
  signal mode properties:
   impulse response, 170
   transfer function, 171–172
  small coupling, 15, 107
Coupled power equations:
 random coupling:
  examples:
   exponential covariance, 127
   white spectrum, 53–58, 71–73
  multi-mode:
   almost-white spectra, 106
   general spectra, 115, 120–121
   white spectrum, 34, 67
  nondegenerate, 79–81
  two modes:
   almost-white spectra, 103
   general spectra, 110, 117–118
   white spectrum, 32
  random propagation parameters, 93–94
   example, 94–96
Coupling coefficients, 6. *See also*
  Random coupling coefficients;
  Square-wave coupling
 multi-mode, 15, 23
 constant, 17
 discrete, delta-function, 16–17

231

232 INDEX

Coupling coefficients *(continued)*
  example, 68
  two modes, 8, 23
    constant, 9–10, 88
    discrete, delta-function, 9–12
Cross-powers, *see* Powers and cross-powers

Dielectric slab waveguide, 67–68, 124, 131, 197–198
  example, 200–202
Differential and difference equations:
  matrix, 212–213
  scalar, 211
Directional couplers, 1–2, 5. *See also* Random propagation parameters

Filter:
  intensity impulse response, 45, 195
  wide-sense stationary transfer function, 44, 191
    correlation bandwidth, 45, 194
    spectral density, 44–45, 192

Impulse response, 25, 196
  statistics, 45–46
  example, 64–66
Inputs:
  coherent, 30, 50–51
  incoherent, 30, 50–52

Kronecker products, 3, 30, 33, 40, 42, 84, 96, 99, 121, 147, 158, 161
  properties, 175–177

MAPLE, 3, 6, 34–35, 37, 47, 49, 63, 81, 143, 154, 162
Marcuse, 74, 80
Markov:
  coupling, 135, 138–139, 218–219, 224
  matrices, 183, 186
  scalars, 183–184
Matrix products, *see also* Kronecker products
  independent matrices:
    expected values, 181
    second moments, 182

Markov matrices:
  example, 189
  expected values, 187
  second moments, 188–189
Modes, 1, 5, 6. *See also* Propagation parameters
  backward, 1, 8, 87
  degenerate, 8, 16
  delay, 24–25
  dispersion, 25
  forward, 1, 6, 8
  nondegenerate, 30, 43, 49, 74
  powers. *See also* Coupled power equations; Cross-powers; Power fluctuations
    multi-mode, 15,16
      conservation, 16
    two modes, 7
      conservation, 7
    signal and spurious, 7, 23–24
  velocities, 24, 43
Multi-layer coatings, 1, 2, 146
  Kronecker products, 151–152
  matrix representation, 146–148, 229
  reflectance and transmittance, 145, 149
  reflection and transmission, 145, 147–148
  series expansions, 149–150
  thirteen-layer filter:
    design transmittance, 154
    parameters, 153
    transmission correlation coefficient, 157–158
    transmission variance, 157
    transmittance, 156
Multi-mode:
  optical fibers, 1, 5
  waveguides, 1, 5,

Optical coatings, *see* Multi-layer coatings

Personick, 66
Perturbation theory, 143, 147, 158, 161
  multi-mode, 18–19, 108
  two modes, 13–15, 106–107
Power fluctuations, 36, 58–59
  multi-mode, 37

nondegenerate, 83
  two modes, 37–38
  examples, 60–63
Powers and cross-powers, 31
  multi-mode, 33–34, 50–52
  two modes, 32, 93–95
Propagation parameters, 6. *See also*
    Random propagation parameters
  multi-mode, 15
    attenuation and phase, 18
    constant, 17, 22, 100
    example, 68
    degenerate, 16, 206
  two modes, 6–7
    attenuation and phase, 6–7, 11–12, 22, 24–26, 38, 40
    constant, 9–10, 22, 30, 49, 100
    degenerate, 8, 203

Random coupling coefficients, *see also*
    Square-wave coupling
  almost-white spectra, 100–101
  correlation length, 13, 101
  Gaussian or Poisson, 21
  general spectra:
    band-pass, 107
    low-pass, 100
  white spectrum, 21, 31, 49
Random optical thickness, 2. *See also*
    Multi-layer coatings
Random parameters:
  coupling, *see* Random coupling coefficients
  guide width, 2
  layer thickness, *see* Multi-layer coatings
  optical thickness, *see* Multi-layer coatings

propagation, *see* Random propagation parameters
separation between guides, 2
straightness, 2, 5
Random propagation parameters, 88–91
  correlation length, 89

Schelkunoff, 5
Square-wave coupling:
  coupling function, 130, 215
  coupling spectra:
    binary independent, 133, 218
    binary Markov, 136–137, 218
    five-level, band-pass Markov, 141, 224
    six-level, low-pass Markov, 140, 219
  examples:
    four modes:
      six-level, low-pass Markov sections, 141–142
    two modes:
      binary independent sections, 132–134
      binary Markov sections, 135, 138
  inputs, 130

Transfer function(s), 23–24
  covariance:
    multi-mode, 43
    two modes, 42
Transmission statistics, 1, 2, 6. *See also*
    Average transfer functions;
    Coupled power equations; Impulse
    response, statistics; Power
    fluctuations; Square-wave
    coupling; Transfer functions,
    covariance; Multi-layer coatings